Fuel/Engine Interactions

Other SAE books of interest:

Future Automotive Fuels and Energy Technology Profile
(Product Code: T-128)

Alternative Fuels: Emissions, Economics, and Performance
(Product Code: R-143)

Internal Combustion Engine Handbook
(Product Code: R-345)

Fuel/Engine Interactions

Gautam Kalghatgi

SAE INTERNATIONAL

Warrendale, Pennsylvania
USA

Sponsored by

ارامكو السعودية
Saudi Aramco

SAE INTERNATIONAL

400 Commonwealth Drive
Warrendale, PA 15096-0001 USA

E-mail: CustomerService@sae.org
Phone: 877-606-7323 (*inside USA and Canada*)
724-776-4970 (*outside USA*)
Fax: 724-776-0790

This edition is made possible in part by sponsorship of Saudi Aramco.

ارامكو السعودية
Saudi Aramco

ISBN 978-0-7680-6458-2
SAE Order Number R-409
DOI 10.4271/R-409

Library of Congress Cataloging-in-Publication Data
Kalghatgi, G. T.
Fuel/engine interactions / Gautam Tavanappa Kalghatgi.
pages cm
Includes bibliographical references and index.
ISBN 978-0-7680-6458-2
1. Internal combustion engines—Design and construction. 2. Motor fuels.
I. Title.
TJ789.K27 2013
629.25'38—dc23
2013022940

To purchase bulk quantities, please contact:
SAE Customer Service
Email: CustomerService@sae.org
Phone: 877-606-7323 (*inside USA and Canada*)
724-776-4970 (*outside USA*)
Fax: 724-776-0790

Visit the SAE International Bookstore at
http://books.sae.org

Dedicated to the memory of my father, Prof. T.G. Kalghatgi

Table of Contents

Preface

The energy scenario, of which transport is but one sector, is shaped by political and economic factors, such as energy security and local environmental concerns, and global issues, such as climate change. World demand for energy in general, and for transport in particular, is increasing rapidly. Moreover, the increase in demand for transport energy is expected to be skewed heavily toward commercial transport. The internal combustion (IC) engine, fueled mostly by petroleum-derived liquid fuels, has been the main source of transport power over the past century and is likely to remain so in the foreseeable future, though alternatives such as electrification of the transport system will play a role. However, engines are changing, primarily driven by the need to be more efficient. They will also have to meet increasingly stringent emissions standards, be affordable, and satisfy customer demands such as driveability in a globally competitive market. Fuels will continue to evolve to minimize local and global environmental impact in their manufacture and in their use in engines. Conventional fossil fuels will constitute most automotive fuels in the foreseeable future, but they will have to change to adapt to changes in engine technology. Unconventional transport fuels such as biofuels, gas-to-liquid (GTL) fuels, compressed natural gas (CNG) and liquefied petroleum gas (LPG) will also have a role to play. Hydrogen might become a viable transport fuel if it overcomes significant barriers in production, transport, storage, and safety. Also, the fuel cells that require hydrogen must become viable.

This book considers these issues briefly in Chapter 1, and Chapter 2 introduces practical transport fuels. Chapter 3 discusses an important practical topic about a fuel effect on engines that is not usually considered in text books—deposits in engines. However, the primary focus of the book comprises the implications of IC engine development trends for future fuels. In particular, the autoignition behavior of practical fuels provides the unifying thread to help understand how these changes might influence future fuels. An attempt is made to bring together fundamental studies in autoignition such as ignition delay measurements in shock tubes and experimental studies in IC engines.

Measures aimed at increasing the efficiency of spark ignition (SI) engines such as downsizing and turbocharging lead to an increase in the pressure inside the cylinder for a given temperature. This will make knock, an abnormal combustion phenomenon that limits efficiency, more likely. The fuel needs to have sufficient antiknock quality, measured by a combination of research and motor octane numbers, RON and MON, to enable such engines to run at their designed efficiencies. The RON and MON scales are based on primary reference fuels (PRFs) whose autoignition chemistry is very different from that of practical fuels. As pressure increases for a given temperature in the cylinder, practical fuels become more resistant to knock compared to PRFs. The result is that historically since the 1930s, as engines have become more efficient, the importance of MON has declined. Indeed, in most modern cars for a given RON, a *lower* MON fuel has more resistance to knock. However, current fuel specifications assume high MON to be important for knock resistance of the fuels. Thus in the United States, antiknock quality is defined by (RON + MON)/2, while in Europe there is a minimum MON specification of 85. This already has had a distorting effect in the sense that refineries are spending more money and energy in making fuels that are arguably less suited for modern engines. This mismatch between fuel specifications and engine requirements will get more marked as engines become more efficient, and fuel specifications will need to be changed to reflect the appetites of modern engines.

In HCCI (homogeneous charge compression ignition) engines, fuel and air are fully premixed (as in SI engines), and energy release occurs through autoignition, which causes knock in SI engines. HCCI engines can be run in pressure and temperature regimes not normally accessible to SI combustion, and the study of HCCI engines has greatly helped in increasing our understanding of fuel effects on autoignition. Future turbocharged SI engines will also be more susceptible to preignition, which could then lead to extremely violent knock, called super-knock. Chapters 4 and 5 address these issues.

In diesel engines NOx can be reduced by low-temperature combustion, and soot can be reduced by promoting more premixed combustion. If soot formation is avoided, high levels of exhaust gas recirculation (EGR) can be used to control NOx without increasing engine-out soot emissions. However, diesel fuel ignites soon after injection into the cylinder, making it very difficult to get premixed combustion. Indeed, much of the technology that makes diesel engines expensive and complicated (e.g., high injection pressures) is aimed at overcoming the difficulty presented by the low ignition delay of the diesel fuel. In addition, complex and expensive exhaust aftertreatment systems are required to reduce NOx and particulates to acceptable levels. Recent work has shown that gasoline-like fuels that are resistant to autoignition make it significantly easier to achieve low-NOx, low-soot, compression ignition (CI) combustion.

Moreover, compared to today's gasolines, the optimum fuel would have much lower octane number (RON 70–85) and be less processed, and therefore it would need lower energy to produce compared to conventional gasoline or diesel. Developing the technology to use lighter fractions like naphtha in CI engines will also help match future fuel demand to fuel supply. Otherwise, such light components will be in surplus as the fuel demand is expected to move decisively toward heavier fuels. In CI engines using such fuels, the injection system would have a low injection pressure, comparable to that of a direct injection SI engine, and the aftertreatment focus would be on HC and CO reduction rather than NOx reduction. The engine would be at least as efficient as current diesel engines. Thus future high-efficiency CI engines could be simpler and cheaper than current diesel engines and prefer fuels that require less energy to make than current fuels. However, significant development work is needed to overcome practical problems such as cold start and optimizing subsystems such as injection equipment, EGR and turbocharging (Chapter 6).

Chapter 7 discusses the implications of such engine technology developments for future fuels. In a likely long-term scenario, there would be a minority (40%?) of SI engines and a majority (60%?) of CI engines because they are more efficient than SI engines. Efficient SI engines will require high RON and low MON fuels. Components like ethanol will play an important role in enabling such fuels. Gasoline fuel specifications would need to be brought in line with the requirements of modern engines. Most of the fuel would be for the more efficient CI engines, but would have a RON of 70–85 with no stringent requirements on fuel volatility. There would be little requirement for today's diesel fuels. Common fuel components could be used for both SI and CI engines increasing fuel blending flexibility. Current biodiesels with their high cetane numbers would be less attractive. The high cetane number of current GTL will not have a premium. The heavier components in the oil barrel would need to be cracked but not upgraded.

It is assumed that the reader of this book is familiar with basic concepts of IC engines. The material here has been used in lectures for graduate courses in some European universities. A more comprehensive graduate course called "Fuel-Engine Interactions" included all the topics discussed in this book except for the chapter on preignition and super-knock. The students in this course were evaluated on the basis of a report they submitted and a presentation they made on a topic in the broad area of fuel use in IC engines.

Finally, I would like to thank all my colleagues, coworkers, friends, and mentors in Shell and in different universities who have enabled me to work in this very interesting area. I would also like to thank colleagues and friends who have reviewed and commented on some of the chapters, as well as my current employers at Saudi Aramco, who have encouraged me and supported me in writing this book.

Chapter 1
Introduction: Outlook for Energy and Transport Fuels

Transport is of central importance to modern industrial society. The world has more than a billion motor vehicles [1.1], and the transportation sector is a major segment of many developed economies. Transportation products and services account for around 20% of the dollars spent in the United States, and 1 in 16 Americans is employed by this sector [1.2], and that number goes up to 1 in 7 if indirect employment is included. On a worldwide basis, transport accounted for about 20% of the total energy consumed in 2011 [1.3].

Many organizations publish detailed information and forecasts about energy, including transport energy [1.3–1.8]. Some take a scenario approach, with different assumptions being made for government policies under each scenario but with some common assumptions about population growth and macroeconomic conditions [1.3, 1.6, 1.8]. Detailed projections by fuel, energy sector, and region are then made to identify trends in energy demand and supply and for the consequences of these trends, like the effects on local and global (CO_2) emissions. Government policy aimed at ensuring energy security and tackling local pollution and global climate change will naturally affect these trends.

Many of these assumptions are based only on judgments about market factors such as future rates of growth, technology improvements, and government policies over the decades ahead. The various institutions that publish these forecasts make different assumptions, and their forecasts cover different time periods: The BP outlook [1.5] forecasts through the year 2030; IEA (International Energy Agency) [1.3] and U.S. EIA (Energy Information Administration) [1.7], 2035; ExxonMobil [1.4], 2040; and Shell [1.6] and World

Energy Council (WEC) [1.8], 2050. Projections are highly dependent on the data, methodologies, model structures, and assumptions used in their development and are not statements of what will happen, but rather what might happen under the assumptions made.

It is also prudent to remember, as pointed out by Smil [1.9, 1.10], that forecasts on energy have often proved to be wildly inaccurate especially when they have assumed very fast energy transitions and predicted radical departures from the status quo. For instance, predictions from respected commentators in the 1970s and 1980s that nuclear or renewable energy would capture a large share of the energy market by now have failed to materialize. Variations in the quantitative forecasts are only to be expected from these different sources, but there is remarkable agreement on many core findings and trends. Apart from the projections, statistical data on energy recording historical reality can be downloaded from several sources [1.11, 1.12].

1.1 Global Energy Demand

Energy demand is primarily shaped by population and prosperity. World population is expected to increase to about 9 billion by 2050 from the current 7 billion [1.13], with most of this growth occurring in developing countries—mostly India and Africa. Moreover, in these regions the working age population, aged between 15 and 65 years, will also grow rapidly. This will lead to strong GDP (gross domestic product) growth [1.4] and increased per capita use of energy as millions of people escape poverty. Historically, per capita energy use has always increased with per capita GDP as different countries have developed. However, historic data suggest that once a certain level of GDP is attained, per capita energy use does not increase much with further GDP increase (e.g., *Shell Energy Scenarios to 2050*, p. 29 [1.6]). It would appear that cheap energy is a prerequisite for the economic development of the poorer parts of the world.

World energy demand is expected to increase in the coming decades, with the growth almost entirely outside the OECD (Organisation for Economic Co-operation and Development) countries, as illustrated in Fig. 1.1, which is a projection from the EIA [1.7].

These estimates, like some others (e.g., [1.5]), are for traded primary energy for which reliable statistics exist. Primary energy refers to energy at the source that has not been through any conversion process. In many developing countries nontraded fuels, such as combustible renewables like wood and cattle dung, are still important, particularly in domestic use. The IEA estimated that around 2.7 billion people relied on such biomass for cooking [1.14], and 59% of domestic energy consumption in non-OECD countries in 2005 was from such

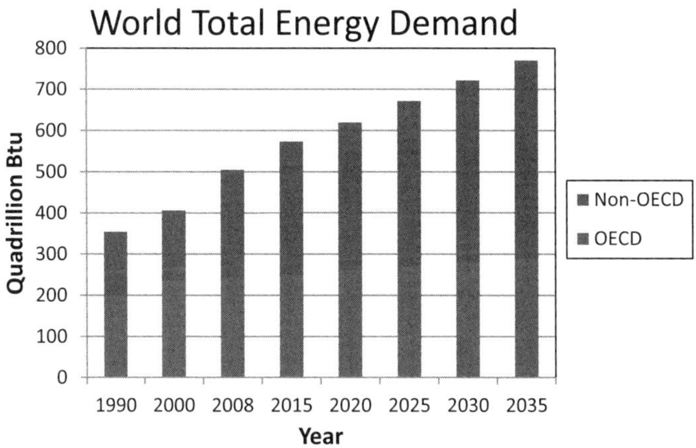

Note: 1 quadrillion Btu = 172 million barrels of oil equivalent (boe) or 23.6 million metric tons of oil equivalent (Mtoe).

Fig. 1.1. World energy demand, EIA [1.7]. Data after 2008 are based on projections, and most of the increase is projected to be in non-OECD countries.

renewables [1.15]. However, this sector is likely to be a small fraction of the total world energy consumption and is diminishing. The projected growth in primary energy demand in Fig. 1.1 between 2009 and 2035 is around 50% and is higher than the approximately 35% growth predicted by ExxonMobil [1.4] or the approximately 40% growth predicted by IEA [1.3] in its central "New Policies" scenario for the same period. However, the energy demand growth rate of around 1.6% per annum projected by the EIA [1.7] is very similar to that projected by BP [1.5].

The *energy intensity,* defined as the amount of energy used per unit of GDP, has been decreasing steadily worldwide, by an average of 1.4% per year between 1990 and 2008; since 2004 that rate has increased to 1.9% per year [1.5, 1.15, 1.16]. All the projections for future energy demand assume that such improvements in energy use efficiency will continue in the future. These efficiency gains are expected to come from improvements in technologies and energy management practices. Without such efficiency gains, the growth in demand will be very significantly higher.

Electricity generation will be the fastest growing source for energy demand in the future [1.4, 1.5], and most of this growth will again be in non-OECD countries. In 2008, of the total energy consumed in the world, around 50% was in the industrial (energy, manufacturing, and chemical) sector, 30% was in the residential and commercial sector, and 20% was in the transport sector

[1.17]. These shares are not expected to change significantly over the coming decades [1.3].

The global energy mix is shaped by the evolution of technology, and the pace of change in energy systems is relatively slow [1.9, 1.10], on the order of many decades. Until the 19th century, the primary sources of energy were renewables in the form of biomass (mostly wood) and human and animal muscle power. Indeed, this is still the reality for billions of poor people in the world who have little access to modern energy systems. With the invention of the steam engine, electricity, and the internal combustion engine, and as industrialization progressed, first coal and then oil entered the energy mix. The big expansion in natural gas only happened after World War II, enabled, according to Smil [1.9], by technology such as new steel alloys, better welding, better pipe-laying, and new compressors that enabled gas to be transported easily using pipelines.

The historical and projected energy supplied by different sources is shown in Fig. 1.2, with data from [1.7]. Figure 1.2a shows the consumption for each source, and Fig. 1.2b shows the share of each source as a percentage of the total energy consumed.

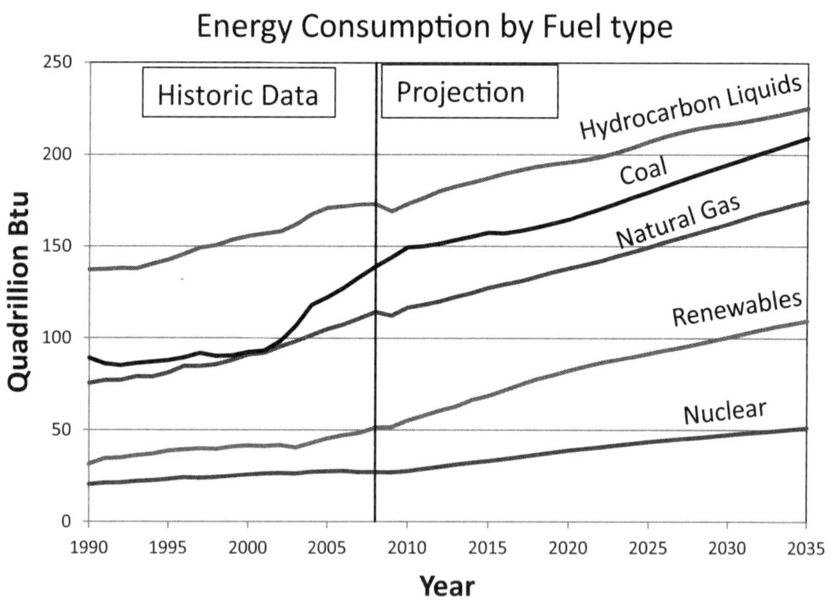

Fig. 1.2a. World energy consumption by fuel type. (Data from EIA [1.7])

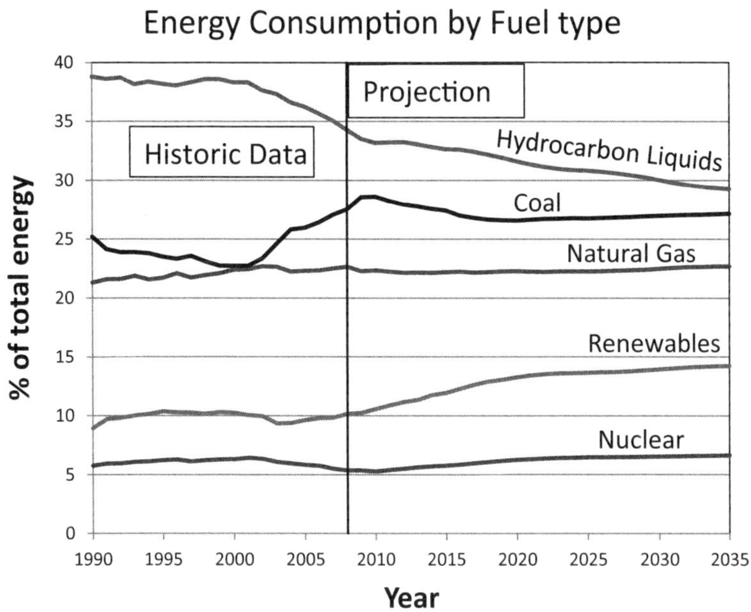

Fig. 1.2b. Share of each source as a percentage of the total world energy consumed. (Data from EIA [1.7])

The share of natural gas might be higher in the future than projected in Fig. 1.2b if cheap shale gas becomes abundant worldwide (see Section 1.2.1). In 2008 around 84% of the world's energy needs were met by fossil fuels: coal, oil, and natural gas [1.7]. This share is expected to decrease only slightly to around 79% by 2035, in spite of the growth of alternatives; other projections come to similar conclusions. Hence, CO_2 emissions, which are of concern because of concerns about global climate change, are expected to grow during the coming decades at almost the same rates as in the recent past. The rate of increase of CO_2 emissions from OECD countries might decrease in the future as energy demand moderates, governments impose more stringent requirements to improve fuel efficiency and reduce CO_2 emissions, and as the share of renewables increases. However, this is likely to be offset by the non-OECD countries, where energy demand is likely to increase rapidly and the poorest people move away from renewables for their primary energy needs.

Table 1.1 shows the conversion factors for different large units of energy used in the different projections.

Table 1.1 Energy Conversion factors

	Joule	kcal	Barrel of Oil Equivalent (boe)	Metric Tons of Oil Equivalent (Mtoe)	Cubic Meters of Natural gas	Metric Tons of Coal Equivalent (Mtce)	Tera-watt Hour
1 quad	1.055×10^{18}	2.52×10^{14}	1.724×10^{8}	23.57×10^{6}	28.32×10^{9}	36.02×10^{6}	293

Note: 1 quad = 1 quadrillion (10^{15}) Btu, 1 exajoule = 10^{18} joules, 1 barrel = 159 liters = 42 U.S. gallons.

1.2 Global Energy Resources and Supply

The important questions that face the world today are whether the increasing global demand for energy can be met and at what financial and environmental cost. Inevitably, there are widely differing views, particularly about environmental costs. For instance, concern about greenhouse gas emissions and climate change has led to policy initiatives in many countries to shift the balance away from fossil hydrocarbon fuels toward renewables. There are arguments from environmental groups that the cost of the energy source should reflect the true environmental costs—hence the demands for a carbon tax. Such a tax would make all energy more expensive than it could be in the short and medium term.

If past experience is anything to go by, affordable energy is a prerequisite for economic growth in poor countries, though there are arguments advanced in industrially developed societies that future growth is more likely to be stimulated by moving to new technologies based on new energy systems. However, in the global debates on energy and climate change so far, developing economies like India and China that want to lift their populations toward prosperity have been particularly unwilling to give up the opportunity afforded by cheap energy. The energy debate is extremely complex, and there are often passionate advocates for one panacea or the other. In reality, there are unlikely to be magic bullets. History tells us that transitions in energy systems span generations, and changes are evolutionary [1.9, 1.10]; thus there will be a mixture of solutions in the future.

We will now briefly consider the supply prospects for different energy systems. In this discussion, the term *reserves* refers to known quantities of resources that are deemed economically recoverable with existing technology.

1.2.1 Fossil Fuels

Around 80% of the world's energy needs are now supplied by conventional fossil fuels and include conventional oil, natural gas, and coal produced using well-established methods. These fuels are relatively easy to recover and are usually of high quality. However, known resources of unconventional fossil fuels, which include oil sands, shale oil, heavy oil, shale gas, and coal-bed methane (CBM) are far larger than conventional fossil fuel reserves [1.18, 1.19, 1.20]. Special recovery techniques have to be employed to exploit such unconventional resources, and further processing is needed to improve their quality. Their use becomes economically viable if the price of energy is high enough, and investment in their development can only be sustained if the price of energy is expected to remain high. Such enhanced recovery techniques are being developed, and unconventional resources are increasingly coming into play.

Many of the technologies developed such as horizontal drilling and hydraulic fracturing (fracking) used for enhanced gas recovery could also be used to significantly increase production from existing oil fields [1.21]. In addition, methane trapped as hydrate in marine sediments in the deep ocean and in ice is another immense resource [1.22]. Currently, no technology exists to exploit this vast energy resource, and great risks are associated with any recovery. For example, methane is many times more potent than CO_2 as a greenhouse gas, and any technology that might be developed to exploit gas hydrates must ensure that no accidental, large-scale release of methane to the atmosphere would occur.

Figures 1.3 and 1.4 show the evolution of oil and gas reserves between 1980 and 2010.

Also plotted in each of these figures is the ratio of the reserves to the annual production level. Over three decades, the reserves increased steadily and more than kept pace with increasing demand. Thus, in 1980 there was enough oil to last for around 29 years at the production level of that time, while by the end of 2010 there was enough oil to last around 46 years. During these three decades, each year there was enough natural gas to last for more than 60 years. The EIA data [1.12] paint a similar picture—in 2009 the world oil reserves would have lasted for about 44 years at the then-annual production level of about 84 million barrels per day.

Oil and gas reserves have gone up because each year there are new discoveries and the recovery rates have been improving. It is impossible to recover and produce all the oil in a known oil deposit; for example, in 2007 the average worldwide recovery factor (i.e., the fraction of the in-place oil discovered that is technically recoverable) was estimated to be only 22% [1.23]. Oil recovery factors can vary very widely, and the best recovery factors could approach 70%, though such efficient recovery is rare [1.23]. However, enhanced oil recovery (EOR) techniques, such as injecting fluids or gas (e.g.,

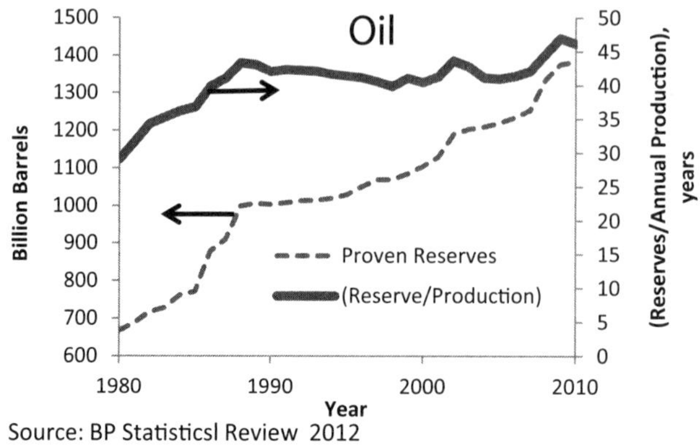

Fig. 1.3. Proven oil reserves and reserves per annual production ratio. In 2010, reserves would have lasted for about 46 years. (Source: BP Statistical Review [1.11])

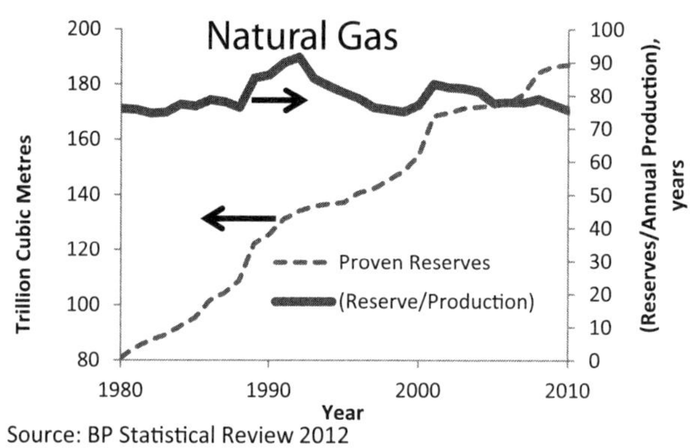

Fig. 1.4. Proven natural gas reserves and reserves per annual production ratio. The reserves would last for around 75 years at current rates of production. (Source: BP Statistical Review [1.11])

CO_2) to force oil out, can significantly improve recovery rates. For instance, in the United States the average recovery rate improved from about 22% in 1979 to about 39% in 2007 [1.23]. Such EOR techniques become feasible when the oil price is high and clearly can have a huge impact on reserves. Contrary to popular belief about "peak oil," oil supply capacity is growing worldwide at an unprecedented level, and there is an abundance of oil to meet future demands [1.10, 1.21].

Coal is the most widely available fossil fuel, and large quantities of coal are available in the world [1.20, 1.24]; current reserves are expected to last for about 120 years at current consumption rates. More than 40% of the world's electricity is generated from coal, and this share is expected to increase in the coming decades [1.20]. China and India, where the electricity demand is expected to increase most rapidly, also possess abundant reserves of coal, and they now rely heavily on coal for electricity. China and India produce about 80% and 70%, respectively, of their electricity from coal [1.20].

Of all the fossil fuels, coal combustion produces the most (and natural gas the least) CO_2 per unit energy. Hence, many Western governments are trying to discourage the use of coal because of concerns about climate change, and some require that new coal-fired power stations only be built if they incorporate carbon capture and storage (CCS). CCS technology separates the CO_2 from the other products of combustion of fossil fuels; it is then compressed and transported via pipelines and stored safely in deep underground structures, such as depleted oil and gas reservoirs and deep saline aquifers [1.25, 1.26]. Some estimates suggest that CO_2 capture will reduce the efficiency of a powerplant by about 15% [1.27]; additional energy will be required to compress, transport, and store it. This technology is being developed [1.25], but the immense effort and cost needed [1.26] make the widespread, worldwide deployment of CCS unlikely anytime soon.

Unconventional resources are expected to play an increasingly important role in the coming decades. Among the unconventional fossil fuels already making a mark on the energy supply are oil sands, shale gas, and CBM. Exploitation of these resources will always have environmental impacts that need to be addressed.

Oil sands are mixtures of sand, clay, and water saturated with a dense and extremely viscous form of petroleum technically referred to as bitumen or tar. The bitumen in oil/tar sands cannot be pumped from the ground in its natural state. Instead, oil sand deposits are mined, generally using strip mining or open-pit techniques, or the oil is extracted by underground heating with additional upgrading [1.28]. A related unconventional oil resource is heavy or extra-heavy oil, which is very viscous and does not flow as easily as conventional oil [1.20, 1.28]. Taken together, these two oil resources are immense—the WEC estimated that there are more than 5 trillion barrels of this unconventional oil in place across the world, about four times the known reserves of conventional oil (Tables 4.1 and 4.2 in [1.20]). However, of these resources at the end of 2008, WEC classed around 240 billion barrels of bitumen (70% of it in Canada) and around 59 billion barrels of extra-heavy oil, (almost all of it in Venezuela) as reserves (i.e., recoverable resources).

In addition, another vast, and as yet untapped, oil resource exists in the form of shale oil. The term *oil shale* generally refers to any sedimentary rock that contains solid organic matter called *kerogen*, which is released in the form of petroleum-like liquids when the rock is heated. WEC estimates the worldwide resource for shale oil at 4.8 trillion barrels [1.20]. However, only a very small amount of shale oil, around 18 thousand barrels per day, is now being produced across the world because of the significant costs of mining oil shale and extracting the oil from the rock [1.20]. If and when demand for oil increases and the price is high enough, more of these unconventional resources will come into play.

The world energy supply is in the process of being transformed by the availability of vast amounts of shale gas [1.29, 1.30], another nonconventional fossil fuel. Sedimentary rock such as shale has low permeability, and gas, mostly methane formed from organic matter laid down during the formation of the rock, is trapped in the pores. Horizontal drilling and hydraulic fracturing (fracking) has enabled gas from such rock formation to be recovered. Further details and the history of the development of this process can be found in other sources [1.29–1.31].

The EIA estimated the technically recoverable shale gas quantity across the world at around 188 trillion cubic meters in 2011 [1.30], and this is the same size as the current conventional reserves of natural gas (Fig. 1.4). The United States has led the world in exploiting shale gas: In 2010 shale gas production was around a third of the total natural gas production in the United States, which represents an increase by a factor of 11 over the preceding decade [1.32].

Coal-bed methane (CBM), also known as coal seam gas in Australia, is methane adsorbed on the internal surfaces of pores inside coal. It is held in place by hydrostatic pressure from water that permeates coal beds. To produce methane from coal beds, this water must be drawn off first, lowering the pressure so methane can flow out of the coal [1.33–1.35]. CBM accounts for about 7% of all natural gas production in the United States and is set to become an important source of energy in Australia [1.35]. An assessment in 2007 by the NPC (National Petroleum Council) estimated worldwide CBM resources to be 256 trillion cubic meters [1.36], though it is now being exploited only in a few countries.

Tight gas is natural gas locked in extraordinarily impermeable hard rock, sandstone, or limestone formations that are atypically impermeable or nonporous, also known as tight sand [1.36, 1.37]. The rock has to be cracked using fracking technologies to recover the gas. One estimate [1.36] suggests that the world has around 180 trillion cubic meters of tight gas, though this resource is not really being exploited outside the United States at this time.

The availability of cheap, unconventional gas could also have a significant impact on CO_2 emissions. For instance, the abundance of shale gas led to very low natural gas prices in the United States in 2011 and a noticeable shift from coal to gas in electricity generation [1.38] enabling a reduction in CO_2 emissions. China also has large shale gas resources [1.30], and shale gas is set to have a significant impact on world energy supply in the future. The estimates of reserves, production levels, and the impact on energy use are likely to change in a field that is evolving rapidly.

Table 1.2 lists the worldwide fossil fuel reserves and resources just discussed.

Table 1.2 World Fossil Fuel Reserves

	Reserves	Data Source, Year of Estimate	Years of Reserves at Current Production Rates*	Reserves (in quad)
Conventional oil	1383 billion barrels	BP [1.11], 2010	46	8022
Oil from oil sands	650 billion barrels	WEC [1.8], 2011	22	3770
Heavy/extra-heavy oil	434 billion barrels	WEC [1.8], 2011	14.5	342
Conventional natural gas	187 trillion m^3	BP [1.11], 2010	58.5	6603
Shale gas	188 trillion m^3	U.S. EIA [1.30], 2010	59	6638
Coal	861 billion tons	WEC [1.20], 2008	128	23,900
				Total 46,897

*Oil, ~30 billion barrels Pa; gas ~3.2 trillion m^3 Pa [1.11]

Additional Unconventional Resources

It is not certain how much is recoverable.

Shale oil	4.8 trillion barrels	WEC [1.20] / 2008
Coal-bed methane (CBM)	256 trillion m^3	NPC [1.36] / 2007
Tight gas	210 trillion m^3	NPC [1.36] / 2007
Gas hydrates	2800 trillion m^3	CRS [1.22] / 2010

Notes:
1. 1 quad = 1 quadrillion Btu; see Table 1.1 for conversions.
2. Reserves are resources that can be recovered economically using current technology. They have been increasing every year with new discoveries and improved recovery rates.
3. World annual demand for energy, including renewables and nuclear, was 506 quad in 2008 [1.7].
4. Fossil fuel energy production amounted to 426 quad in 2008 [1.7]

1.2.2 Renewable Energy

Renewable energy sources include solar, wind, hydropower, geothermal, and biomass. Unlike fossil fuels, their supply is not, in theory, limited, and their direct use does not cause a net increase in CO_2 emissions. Governments across the world are promoting renewables with many policy initiatives to improve energy security, diversify the energy mix, and reduce CO_2 emissions. Most renewables are used to produce electricity, though many poor people burn combustible renewables like wood for their primary energy needs. The CO_2 produced from burning biomass is a part of the carbon cycle and does not increase the concentration of CO_2 in the atmosphere, because it is absorbed by new plant growth. There has been a rapid increase in electricity generation using renewables. For instance, global wind power capacity has increased from around 18 gigawatts (GW) in 2000 to 238 GW by the end of 2011, and solar photovoltaic (PV) capacity, which converts solar energy directly to electricity, has grown from around 1.5 GW to 67 GW over the same period [1.17]. The annual growth rate for the five years leading up to 2011 has been around 25% for wind and around 50% for solar PV. This rapid growth is expected to continue, and the share of renewables is projected to increase from around 10% in 2008 to 14% of total energy in 2035 (see Fig. 1.2b).

In theory, all the energy needs of the world could be supplied by renewables given enough time and investment, but there are several issues associated with electricity generation from renewables that need to be resolved before they can substantially replace fossil fuels. Most renewables are intermittent. Wind does not blow all the time, and the sun does not shine on the solar panels at night or on cloudy days. In actual operation, the load factor, the energy delivered on average as a percentage of the maximum installed capacity, has been 20–25% for wind and around 16% for solar PV compared to an average of 74% for U.S. coal-fired powerplants [1.9, 1.26]. This means that conventional backup power generation capacity is always needed to cover situations when wind or solar energy fails to deliver, usually when energy is most needed (e.g., on cold winter nights). They are also often diffuse—the capacity of a wind turbine is around two orders of magnitude lower than a conventional thermal plant, so that many wind turbines over a wide geographical area need to be built to make a substantial contribution. This will require extremely large investments in grid and other infrastructure [1.26]. Electricity generation from renewables is usually significantly more expensive than conventional power generation. Figure 1.5 shows the levelized cost of electricity generation from different sources calculated by the EIA [1.39].

The levelized cost takes into account the costs for capital investment, operating and maintenance, and fuel, as well as the capacity or load factor (availability) of each technology. Of the various renewable options for electricity,

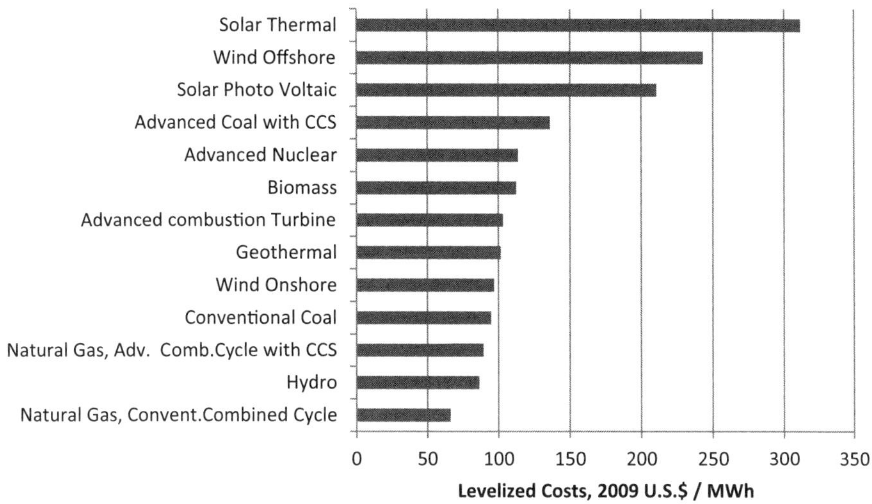

Fig. 1.5. Levelized costs of different electricity generation technologies from EIA [1.39].

hydroelectric generation is the cheapest, though the potential of finding new locations and hence for growth in this area is limited. Solar and offshore wind are the most expensive. In most cases governments have to subsidize renewable energy to make it competitive with cheaper options. As the technology develops, costs of renewables will come down, but the prospects for substantial growth depend on government policies and the costs of coal and gas. Even with a threefold growth in the IEA central scenario [1.17] between 2009 and 2035, the share of renewables increases from 13% to only 18%. There is concern now that the growth in renewables might stall if gas becomes cheaper because of shale gas, though the immediate impact of cheap gas in the United States has been on replacing coal as the energy source for electricity [1.38].

1.2.3 Nuclear Energy

It was once thought that all the world's energy problems would be solved by nuclear energy, but that has not come to pass, because there are unresolved and hotly debated issues about the cost, safety, disposal of nuclear waste, and security surrounding possible theft of fissile material by terrorists. Nuclear energy provides about 5.8% of the primary energy [1.7] and around 13% of electricity to the world. It is expected to grow by about 70% between 2009 and 2035, led by China, Korea, and India, to make up about 7% of the world's primary energy share by 2035 [1.7]. Electricity generation from nuclear energy does not produce any CO_2 and is seen by many as an essential element in mitigating CO_2-driven climate change.

1.2.4 Efficiency Improvements

Efficient use of energy will be an important enabler for reducing the demand for energy, controlling emissions, and increasing energy security for individual countries. Energy intensity (ratio of per capita energy use to per capita GDP), on a worldwide basis has been decreasing steadily, by an average of 1.4% per annum between 1990 and 2008 [1.5, 1.15, 1.16]. An exception to this trend is the Middle East, where energy intensity has been increasing [1.16]. Without such improvements, the world primary energy consumption from 1973 to 2005 would have been higher than it actually was by 58% [1.15].

There is still great potential for improving efficiency in almost all areas. For example, in India and China, demand for electricity is rising sharply. Currently, India produces around 70% (China, 80%) of the electricity from coal with an average efficiency of only 27% (China, 32%). However, the best-in-class coal-fired power stations have an efficiency of 48% [1.40]. Similarly for steel and cement production and household appliances, the average efficiency is significantly worse than the best-in-class efficiency, especially in emerging economies [1.16]. Economic, fiscal, and legislative incentives from governments will be critical for promoting the efficient use of energy but have to be designed and implemented effectively [1.16].

1.3 Transport Energy and Transport Fuels

Because liquid fuels have high energy density and are easy to transport, store, and handle, they have become the fuel of choice for transport over the past century. As a result, a very extensive worldwide infrastructure has been established for the production and distribution of liquid fuels worth trillions of dollars. At normal temperature and pressure, natural gas has around 800 times less, and hydrogen over 3000 times less, energy per liter compared to gasoline. Hence, when gaseous fuels are used to power transport, they inevitably have to be compressed or liquefied to increase their volumetric energy content. This requires additional energy and infrastructure. Table 1.3 lists some typical properties of fuels used as transport fuels.

1.3.1 Conventional Transport Fuels

A very large fraction of world transport energy comes from petroleum (oil), which is likely to remain the case in spite of all the changes that are expected in the transport sector. According to [1.4], in 2010 around 95% of transport energy was derived from petroleum, and by 2040 this share will decrease to around 90%. The world demand for transport fuels is very large. Table 1.4 shows a snapshot of oil products demand for April 2012 [1.42].

Table 1.3 Some Properties of Some Transport Fuels

Properties	Lower Heating Value MJ/kg	Density kg/liter	Lower Heating Value MJ/liter	Initial boiling point (IBP), °C	Final boiling point (FBP), °C
Gasoline	43.5	0.74	32.1	28	198
Diesel	43.2	0.83	36.0	165	352
Natural gas [1.41]	50	7.35×10^{-3}	0.04		
LNG [1.41]	55	0.45	24.8	-160	
LPG [1.41]	46.2	0.522	24.1	-40	0
Hydrogen	120	8.52×10^{-4}	0.0102		
Liquid hydrogen	120	0.071	8.5	-253	

Notes. Natural gas contains mostly (70–90%) methane, though it can also have CO2, nitrogen, propane, hydrogen sulfide, and ethane, and its composition can vary. LNG = liquefied natural gas, LPG = liquefied petroleum gas, assumed here to contain 70% propane and 30% butane by mass. Both LPG and LNG can vary in composition.

The demand for gasoline, diesel, kerosene or jet fuel, and gas oil is around 54 MMboed (around 114 quad or 121 exajoules (EJ) per year) and constitutes around 60% of the oil products demand. The rest of the oil products go mostly to the chemical industry and heating. The demand for transport fuels is expected to grow rapidly as the number of vehicles increases and, again, because the growth in demand is almost entirely from non-OECD countries [1.3–1.8]. Figure 1.6 illustrates this using data from the EIA.

Table 1.4 A Snapshot of the Demand for Oil Products

	Million barrels of oil equivalent per day (MMboed)		
	OECD	Non-OECD	Total
Total	43.9	43.8	87.7
Gasoline	13.8	8.6	22.4
Diesel or gas oil	11.9	14.0	25.9
Jet fuel or kerosene	3.4	2.7	6.1
Residual fuel oil	2.8	5.6	8.4
Other*	12.0	12.9	24.9

Note. *Other includes naphtha, LPG, and ethane.

Source: Data for April 2012 from IEA. [1.42]

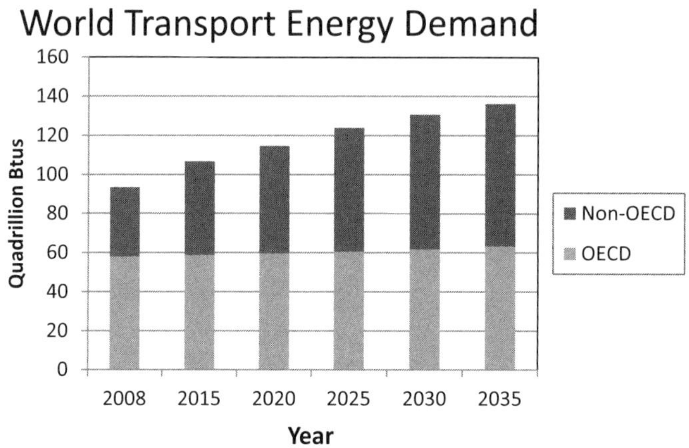

Fig. 1.6. Transport energy demand, according to the EIA [1.7]. The data after 2008 are projections. Demand increases almost entirely because of growth in non-OECD countries.

However, the increase in energy demand is not going to be uniform across all transport sectors [1.4, 1.5, 1.8]. For instance, the number of personal vehicles is expected to double to 1.6 billion units in 2040 compared to 2010, but the fuel demand from this sector is not expected to increase [1.4]. This is because of the expected large improvement in efficiency (e.g., through hybridization; see Section 1.6), reduction in average vehicle size, and reduction in the miles driven compared to today's average vehicle, especially in countries like China and India where the number of vehicles is expected to increase the most. In contrast, the energy demand for the commercial transport sector, which includes heavy-duty trucks, aviation, and marine, is expected to increase by about 75% over the same period in spite of some gains in efficiency [1.4].

According to one projection [1.4], if there are no major changes in engine technology, the demand for diesel fuel, the most popular fuel for heavy-duty vehicles, will rise by 85% in spite of an assumed increase in efficiency of 30%, while gasoline demand will fall by about 10% between 2010 and 2040. The WEC projects a similar mismatch in the demand growth for light and heavy fuels [1.8]. Currently, the ratio of the demand for diesel and jet fuel combined to the demand for gasoline is around 1.4 (see Table 1.4). The WEC projects that this ratio will increase to around 2.4 by 2040 (see Fig. 18 in [1.8]) under the "freeway" scenario where market forces are dominant. Under the more regulated "tollway" scenario of the WEC (see Fig. 24 in [1.8]), this ratio is projected to increase to around 3.8 by 2040. Thus, whatever the actual numbers, fuel demand is going to shift very significantly toward heavier fractions in the future [1.4, 1.5, 1.8], and this is likely to increase the availability of light components like naphtha and straight-run gasoline from the initial distillation of crude oil.

In most parts of the world there are strong ongoing initiatives to reduce the dependence on oil of the transport sector in order to reduce oil imports to increase energy security and improve trade balances. Another goal is to reduce the CO_2 emissions associated with transport to address global climate change concerns. Governments are trying to achieve these goals by setting increasingly demanding fuel efficiency and CO_2 emissions standards for vehicles and by managing the demand for transport energy through other initiatives such as promoting the use of public transport. Reducing the CO_2 emissions from the vehicle via higher efficiency will reduce fuel consumption when hydrocarbon fuels are used.

The CO_2 and fuel economy standards usually require vehicle manufacturers not to exceed some set level measured in a specified test and averaged across all the cars they sell. The regulations also specify penalties if the targets are not met. If a company wants to sell large cars with high fuel consumption and CO_2 emissions, it will have to sell a sufficient number of more fuel-efficient cars to meet the fleet-averaged obligations. For instance, the fleet average to be achieved by all new cars registered in the European Union (EU) is 130 grams of CO_2 per kilometer (g/km) by 2015, a 19% reduction compared with 2006; a further emission reduction to 95 g/km is specified for the year 2020 [1.43]. In addition, controlling local emissions such as soot, nitrogen oxides (NOx), unburned hydrocarbons (HCs), and carbon monoxide (CO) to improve local air quality is a matter of high importance. Local emissions standards are also being tightened across the world to improve local air quality, and ensuring compliance with such standards is increasingly important in non-OECD countries as traffic levels increase. Such measures sometimes conflict with the requirements of reducing fuel consumption, and these conflicts have to be managed. Global emissions and fuel-efficiency standards can be found in different sources (e.g., [1.44]).

Alternative fuels not based on petroleum, such as biofuels, are also being vigorously promoted in many countries. Replacing hydrocarbons with electricity to power vehicles will eliminate tailpipe CO_2 emissions from vehicles and enable vehicle manufacturers to meet tough fleet-averaged CO_2 standards, though the overall impact depends on how electricity is generated. Partial electrification of the powertrain in the form of hybridization improves fuel economy because the internal combustion engine can be run in its more efficient operating regimes. We will briefly consider these alternatives to petroleum-based fuels.

1.3.2 Alternative Transport Energy Sources

We define alternative transport energy sources as those that are not derived from petroleum, which supplies most of the transport energy today. The current demand for transport fuels is very large—the demand listed in Table 1.4 translates to around 1.15 billion tons (~1.55 trillion liters) per year for gasoline and around 1.34 billion tons (~1.6 trillion liters) per year of diesel or gas oil. It will be difficult for alternative transport energy sources to displace petroleum to any substantial extent in the next few decades even if their use grows rapidly.

1.3.2.1 Biofuels

Biofuels [1.45–1.49] are commonly made from sugar and starch from crops such as sugarcane and sugar beet, and corn and vegetable oils. They could also be made from energy crops such as Jatropha, wastes such as waste oils and food-processing wastes, agricultural residues such as straw and corn stover, forestry residues, and novel feed stocks such as algae. Governments across the world promote biofuels for many complex reasons. These reasons include increasing energy security by reducing oil imports, use of agricultural surpluses, rural employment and waste management, and possible greenhouse gas benefits.

The most common biofuels are (1) ethanol made from fermentation and distillation of sugar and starch from sugarcane and corn and (2) biodiesels made from the esterification of vegetable oils. Biofuels are generally oxygenates and have much lower energy content compared to conventional liquid fuels. For example, the energy content of ethanol per liter is only about 60–65% of the energy content of a typical gasoline, whereas a liter of biodiesel contains 85–90% of the energy in a liter of a typical diesel fuel. Many biofuels have desirable combustion characteristics. For instance, ethanol has a very high resistance to knock and could enable higher efficiencies in spark ignition engines—see Chapter 4. The world production of biofuels in 2010 was around 1.9 million barrels per day [1.11]. The global ethanol production in 2011 was around 85 billion liters [1.50], equivalent in energy terms to about 41 million tons of gasoline or less than 4% of global gasoline demand; over 85% of this production was in the United States and Brazil.

Currently, biofuels are produced from crops that can also be used for food and are known as first-generation biofuels. More than 40% of the corn grown in the United States was estimated to have been used for ethanol production [1.45] in 2011. The competition for land between fuel and food is creating increasing concerns [1.45–1.49] and has been blamed for rising food prices. Second-generation biofuels [1.45, 1.51] aim to use nonfood cellulosic biomaterials such as straw, corn stover, and forestry residues. In one process cellulose is first broken down into simple sugars using enzymes followed by fermentation and distillation to produce ethanol [1.45, 1.51]. In another process biomass is converted to liquids—biomass to liquids (BTL)—through a thermochemical route. In the initial gasification process the, bio-material is heated in the presence of steam to produce syngas, a mixture of carbon monoxide and hydrogen. The syngas is then converted to liquid fuels using the Fischer-Tropsch process [1.52].

The early expectations of rapid development of second-generation biofuels have not been fulfilled. In 2007 a U.S. law was passed that would have required 250 million gallons of second-generation biofuels to be marketed in the United States by 2011. This mandate was revised downwards to 6.6 million gallons by the U.S. EPA (Environmental Protection Agency), but in reality there was virtually no second-generation biofuel available in 2011, and fuel producers had to pay a penalty for not using a fuel that did not exist [1.53]. Even in optimistic scenarios [1.51] second-generation biofuels are not expected to make a significant contribution until sometime after 2020.

Any significant penetration of biofuels into the transport sector will make great demands on land use. The IEA has published a road map to achieve a 27% share of transport energy for biofuels by 2050, which will require 100 million hectares [1.48]. To put this in perspective, the entire United States has around 165 million hectares of arable land, which is around 18% of its land mass [1.54]. Such a massive requirement for arable land in a world with increasing population and increasing demand for food is likely to be extremely difficult to manage. However, if virgin land such as rainforest or grassland is converted to grow fuel crops, it could release up to 400 times more CO_2 into the atmosphere than the annual greenhouse gas reduction that the biofuels could provide by replacing fossil fuels [1.49]. There is increasing concern and uncertainty about the environmental benefits of biofuels in light of the new research on the effects of land use change and other issues associated with fertilizer and pesticide use if crops have to be specially grown for fuel [1.45–1.49].

Biofuels cost more to produce than conventional fuels. Consequently, public support is required in most OECD countries to make biofuel production based on agricultural crops economically viable [1.46]. Production costs for ethanol have ranged in recent years from $0.35 to $0.50 per liter of gasoline equivalent (GE) if produced from sugar cane, and from $0.45 to more than $1 per liter of GE if made from grain or sugar beet. Biodiesel based on vegetable oil is produced for about $0.70–1.00 per liter of diesel equivalent (DE). For comparison, gasoline and diesel prices—net of taxes—have ranged between $0.35 and $0.60 per liter in recent years [1.46].

There is significant research being done on producing biofuels from algae [1.55–1.57]—these fuels would not compete with food for land. Microalgae are single-celled, photosynthetic organisms known for their high energy content. In some cases, more than half of that mass consists of lipids, the same material found in vegetable oils. They can also grow very rapidly; some algal strains are capable of doubling their mass several times a day. However, they are very expensive to make; conservative estimates suggest that algal biofuels produced in large volumes with current technology would cost more than $8 per gallon [1.55]. At the time of this writing there were no commercial quantities of algal fuels available, but algae could become an important source of transport fuels in the future particularly if research into genetically modified algae [1.57] bears fruit. The potential and challenges for biofuels from algae are discussed in [1.56, 1.57].

In summary, in 2012 biofuels supplied an average of less than 4% of transport energy worldwide. They are more expensive to produce than conventional fuels. There are increasing concerns about the competition for land between food and fuels. Second-generation biofuels are meant to address this issue, but it is not clear when and to what extent they will contribute. There are also increasing concerns and uncertainties about the environmental benefits of biofuels. Nevertheless, most governments have policies in place to promote the use of biofuels, primarily to reduce oil imports and to some extent support agriculture and rural employment.

1.3.2.2 Gas to Liquids

Syngas, a mixture of carbon monoxide and hydrogen, can be produced from natural gas, coal, or other biomass by heating it in the presence of steam. Syngas is then converted to liquids in the presence of a catalyst using the Fischer-Tropsch process [1.52]. A variety of products—gasoline and diesel, as well as synthetic waxes, lubricants, chemical feeds tocks, methanol, and dimethyl ether (DME)—can be produced using these processes. When coal is the source, the acronym used is CTL, for coal to liquids. When biomass is the source, the acronym used is BTL; the generic acronym of XTL is also used to describe the process. Such fuels are also known as synthetic fuels.

Currently, commercial quantities of high-quality diesel are made by Shell using natural gas [1.58]. South Africa has the only commercial CTL industry [1.59]; gasoline has been produced using coal there since 1955. The ASFE (Alliance for Synthetic Fuels Europe) estimated [1.60] that by 2015 there will be 80,000 boed of synthetic fuels (<0.2% of total fuel demand). Another study by the NPC projects that there could be 0.5 million boed of gas to liquids (GTL) by 2030 but concludes [1.61] that GTL is unlikely to be a major component of global transport fuel supply by 2020.

The capital cost of a GTL plant can be very high, and the economics of GTL production is affected by many factors [1.62]. Moreover, the gas can be burned directly to produce heat or electricity. Hence, GTL makes the most sense when the gas is "commercially stranded"; that is, it has no significant economic value. However, the shale gas revolution that is underway across the world is likely to influence the prospects for GTL. If there is abundant and cheap gas, governments will encourage its use in the transport sector to reduce oil imports. Since GTL can be blended seamlessly into existing liquid fuels, it can use the existing distribution infrastructure and is the most convenient way of bringing natural gas into the whole of the transport sector. Should this happen, it could also help to mitigate the demand imbalance between diesel and gasoline that is expected to occur in the future.

1.3.2.3 Other Alternative Fuels

Combustible gases such as methane, propane, and hydrogen can also be used in spark ignition engines, and fuels such as DME (dimethyl ether) can be used in diesel engines. The storage volume available for fuel is limited on a vehicle, and the main drawback of such fuels is their low volumetric energy content at normal temperatures and atmospheric pressure. Thus, they have always remained as niche fuels, and no infrastructure for their widespread distribution exists on a global basis, unlike for liquid fuels. Their energy density always has to be increased by compression or liquefaction, which costs extra energy and requires further investment in infrastructure. Second, unless they are injected as liquids into the engine, they displace significant amounts of air and reduce the volumetric efficiency of the engine, thereby reducing the maximum power attainable. Even though they have advantages in combustion terms (e.g., higher octane quality) and can reduce emissions from vehicles, gaseous fuels are unlikely to displace liquid fuels to any substantial degree in transport applications globally. Once again, the possibility of cheap and abundant natural gas because of the shale gas revolution in some countries could change this outlook to some extent.

Fuels that are used in the transport sector include CNG (compressed natural gas), LPG (liquefied petroleum gas) and to a limited extent LNG (liquefied natural gas). The technology to use these fuels in cars is well established. Most vehicles that run on CNG are in large fleets that have centralized fueling facilities. Some countries, such as Pakistan and Iran, where more widespread use of CNG exists have developed extensive CNG fueling networks. An estimated 14.8 million vehicles are running on natural gas across the world [1.63]. LPG is a mixture of propane and butane and is a by-product of refining that becomes liquid at moderate pressures. In 2007, an estimated 13 million cars ran on LPG worldwide [1.64].

Hydrogen as a transport fuel has garnered much interest, particularly as a fuel for fuel cells. However, practical fuel cells have not so far materialized. The problems of transport and storage, and hence the required investment in infrastructure, are greater for hydrogen than other gaseous fuels, given its very low molecular weight and, hence, density. More importantly, hydrogen, like electricity, is an energy carrier and has to be manufactured. It can be produced using a wide variety of sources, ranging from water to fossil hydrocarbons, and indeed this has been one of its attractions from an energy security viewpoint. However, these manufacturing processes are very energy intensive, and if the energy used in its production is not from renewable sources, using hydrogen as a fuel for vehicles will, in most cases, increase the overall greenhouse gas emissions [1.65]. Very significant hurdles in terms of production, transport, and storage have to be overcome before hydrogen can become a viable transport fuel.

1.3.3 Electrification

Electrification is another route by which fuel demand for transport can be reduced. The strongest incentives for electrification for vehicle manufacturers are the stringent standards being imposed in many parts of the world on vehicle fleet CO_2 emissions. Such measures are intended to reduce oil imports as well as climate change impacts from CO_2 emissions. In addition, they could help improve local air quality in crowded cities by reducing or eliminating tailpipe emissions.

All electric vehicles have a battery (commonly lithium ion) and a motor to drive the wheels, and most also have an internal combustion engine. The battery does some or all the following: start the engine, run the electric accessories, launch the car, drive the wheels, provide power additional to that of the engine when needed to the wheels, and store unwanted energy for later use. The efficiency of the internal combustion engine varies greatly depending on its operating condition. The electric drive helps manage the flow of energy through the vehicle system to enable the engine to run at or near its most efficient point regardless of the requirements of the vehicle. The battery is the most expensive part of an electric vehicle; its size and cost depend on the degree of electrification. Instead of a battery, a mechanical device such as a flywheel or a compressed air system could be used to store unwanted or otherwise wasted energy for later use. Such systems are supposed to be less expensive than a battery pack, but as yet have not been developed for practical use.

The simplest form of electrification is mild hybridization, which allows the engine to be turned off when the vehicle stops, brakes, or is coasting, but can be restarted quickly. The most common form is the *hybrid electric vehicle* (HEV), such as the Toyota Prius. This approach integrates an internal combustion engine (ICE); an electric motor, including "regenerative braking"; and a battery to manage the energy flow through the system to increase efficiency. During regenerative braking, the electric motor is operated in reverse, acting as a generator. The load from the generator serves as a brake to stop the vehicle while also charging the onboard batteries with the electrical energy that would normally be lost as heat through traditional mechanical friction brakes. The electric motor receiving energy from the battery helps drive the wheels under conditions when the vehicle demands more power than can be supplied by the ICE. This allows the ICE to operate at or near its most efficient condition and reduce fuel consumption. At low loads, when the energy demand from the vehicle is low, the excess energy produced by the ICE can be stored in the battery for later use. However, all the energy used to move the car using HEV technologies comes from the fuel, with no electricity from the grid involved. Hybridization is much easier to implement in relatively small vehicles that do not require excessively large battery packs, and its benefits are most marked when the ICE operates at low loads with frequent stops and starts. Therefore, it is best suited for light passenger vehicles in urban driving, and all future projections assume that HEVs will become widespread and improve fuel efficiency significantly in this sector.

A *plug-in HEV* (PHEV) or *range-extended EV* (REV) such as the General Motors (GM) Volt has a much larger battery that is usually charged by connecting the car to the electric grid. The battery can drive the car over a limited range, (e.g., 40 miles). The car also has a small ICE that can charge the battery as well as extend the range. So, if the car is driven beyond the range allowed by the battery, the energy has to come from the fuel (i.e., the engine alone drives the car if the battery is out of charge). The *battery EVs* (BEV), or pure electric vehicle, has a big battery and no ICE, and all the energy comes from the grid; an example is the Nissan Leaf. However, its range is limited, and it is the costliest of the electric options because of the size and cost of the battery. A PHEV or a BEV has to be necessarily a small car because of the limitation on the size and cost of the battery. Table 1.5 compares different options for electrification to a conventional car.

BEVs and PHEVs are very expensive, at $15,000–$25,000 more than similarly sized conventional cars. For private motorists, the payback time on this premium depends on the relative costs of conventional fuel and electricity and the average distance driven over a year. The Union of Concerned Scientists (UCS) calculated that in the United States, for an annual mileage of 11000 miles, the annual fuel-cost saving of using a BEV (Nissan Leaf) was between $770 and $1220 compared to a gasoline-powered car giving 27 mpg (miles per gallon; 11.5 km/L) with the gasoline price at $3.50 per gallon [1.66]. It should be remembered that most original equipment manufacturers (OEMs) have conventional small cars of comparable size which are much cheaper and with better fuel consumption than 27 mpg (11.5 km/L).

Table 1.5 Different Options for Electrification

	Conventional Car	Hybrid Electric Vehicle (HEV)	Plug-in Hybrid EV (PHEV)	Battery Electric Vehicle (BEV)
Total range	350 miles	450–550 miles	330-370 miles	80–100 miles
Range on battery alone	None	<2 miles	30–70 miles	80–100 miles
Range on gasoline	350 miles	450–500 miles	300 miles	None
Refueling	Fill up at pump	Fill up at pump	Plug-in (daily use); fill up at pump (long trips)	Plug-in
Price premium	Base	$2000-$5000	>$8000	>$12,000

Currently, most OECD governments are subsidizing BEVs and PHEVs to promote their use, but these subsidies will decrease and eventually disappear. By the time this happens, the cost of battery packs needs to come down sufficiently to make such vehicles economically attractive to private customers. Another disadvantage is that the battery-charging times run to many hours compared to the few minutes it takes to fill up a tank at a fuel pump, so a battery-charging infrastructure needs to be built up. Smil pointed out that if electricity were to take a large share of the transport sector, the additional electricity generation required would be substantial, and this additional capacity could not be added in reasonable time [1.10]. A European report came to similar conclusions [1.67].

BEVs and PHEVs are attractive options for auto manufacturers because the car itself produces little or no (for a BEV) CO_2 and allows them to improve their fleet-averaged CO_2 emissions. However, the total level of CO_2 emissions or the total energy consumption on a full cycle analysis depends on the source and efficiency of electricity generation. France, where nuclear energy has a large share of electricity generation, produces much less CO_2 emissions to generate its electricity than China, which burns coal for a large proportion of its electricity generation. Therefore, from an overall CO_2 point of view, BEVs would make more sense in France than in China. According to [1.67], in Europe the overall impact of the introduction of electric vehicles on CO_2 emissions will be slight until 2030 and will, in addition, depend largely on the electricity source. On a lifecycle basis, EVs can offer CO_2 emission reductions only if they use electricity generated from renewable or nuclear sources.

However, in terms of overall energy, BEVs and PHEVs do not offer any benefits compared to conventional vehicles; indeed they might come out worse if the electricity generation has low efficiency, as it does in many countries. Refinery efficiencies in making transport fuels are very high—up to 90% [1.68]. In comparison, electricity generation is very inefficient (see Section 1.5). Smil calculated that assuming an average electricity generation efficiency of 40% (it is actually 27% in India) and after accounting for other losses such as for transmission, in energy utilization terms, a small electric car would be equivalent to a conventional car of 38 mpg [1.10]. The UCS makes a similar calculation for CO_2 emissions [1.66]: In parts of the United States with the most CO_2-intensive electricity generation, a BEV like the Nissan Leaf is equivalent to a conventional car of 33 mpg. Given that there are many conventional cars of comparable size with much better fuel efficiency and significantly lower cost, BEVs and PHEVs will only make sense in energy terms if electricity generation efficiencies are significantly improved.

In summary, widespread use of HEVs will significantly improve fuel efficiency in the passenger car sector. For BEVs and PHEVs to become widely accepted, battery technology will need to improve significantly, the cost will need to come down, and the challenge of building an extensive battery-charging infrastructure as well as meeting increased electricity demand will have to be resolved. In most parts of the world, particularly in emerging economies, where electricity generation is based on coal and is inefficient, BEVs and PHEVs would increase overall energy consumption and CO_2 emissions. In any case, at current prices they are not likely to be affordable—small cars such as the Tata Nano would have comparable energy and CO_2 footprints but cost almost eight to ten times less.

1.4 Summary and Concluding Remarks

Transport energy demand, like all energy demand, is expected to increase significantly over the coming decades, with most of the increase almost exclusively in non-OECD countries. However, this demand increase will be mostly skewed toward commercial transport, including marine and aviation, rather than personal transport because of the differences in growth, the potentials for increasing efficiencies, and changes in usage patterns and vehicle sizes between the two sectors. If ICE technologies do not change significantly, this could mean a massive increase of over 80% in demand for diesel fuel with little change or even a decline in the demand for gasoline in the next three decades [1.4, 1.8], leading to an abundance of light fractions such as straight-run gasoline from the initial distillation of petroleum.

Transport energy is now supplied mostly by liquid fuels made from petroleum. In spite of the many initiatives to reduce dependence on oil, its share is going to decrease only slightly—from 95% in 2010 to 90% in 2040 according to [1.4]. The demand for transport fuels is very large, and even allowing for large growth, alternatives like biofuels cannot substantially displace petroleum from the transport sector. Hybridization will significantly improve the efficiency of the passenger car sector and help mitigate the increase in demand from the increase in the number of vehicles. The scope for the growth of BEVs and PHEVs, which rely on electricity rather than oil for their energy, is limited because of cost, convenience, and the inefficiency and CO_2 intensity of electricity generation.

In the chapters that follow, we will discuss the manufacture and properties of transport fuels, different aspects of their interaction with ICEs, and the implications of engine technology development trends for future transport fuels.

1.5 References

1.1 "World's Vehicle Population Tops 1 Billion Units." Wards Auto, August 15, 2011. http://wardsauto.com/ar/world_vehicle_population_110815. Accessed December 19, 2012.

1.2 "Transportation and Energy Issues." APS. http://www.aps.org/policy/reports/popa-reports/energy/transportation.cfm. Accessed December 19, 2012.

1.3 "World Energy Outlook 2011." International Energy Agency (IEA), 2011. http://www.iea.org/publications/worldenergyoutlook/publications/weo-2011/. Accessed July 4 2012.

1.4 "ExxonMobil 2012 Energy Outlook." ExxonMobil. http://www.exxonmobil.co.uk/corporate/files/news_pub_eo2012.pdf. Accessed May 2, 2013.

1.5 "BP Energy Outlook 2030, January 2012." http://www.bp.com/liveassets/bp_internet/globalbp/STAGING/global_assets/downloads/O/2012_2030_energy_outlook_booklet.pdf. Accessed December 19, 2012.

1.6 "Shell Energy Scenarios to 2050." Shell. http://s09.static-shell.com/content/dam/shell/static/public/downloads/brochures/corporate-pkg/scenarios/shell-energy-scenarios2050.pdf. Accessed January 16, 2013.

1.7 "International Energy Outlook 2011." DOE/EIA-0484(2011), U.S. Energy Information Administration. http://www.eia.gov/forecasts/ieo/pdf/0484(2011).pdf. Accessed December 19, 2012.

1.8 World Energy Council. 2011. *Global Transport Scenarios 2050.* WEC, London.

1.9 Smil, V. 2010. *Energy Transitions: History, Requirements, Prospects.* Praeger, Santa Barbara, CA.

1.10 Smil, V. 2010. *Energy Myths and Realities: Bringing Science to the Energy Policy Debate.* American Enterprise Institute, AEI Press, Washington, DC.

1.11 "Statistical Review of World Energy 2012." BP. http://www.bp.com/sectionbodycopy.do?categoryId=7500&contentId=7068481. Accessed July 7, 2012.

1.12 "International Energy Statistics." U.S. Energy Information Administration. http://www.eia.gov/cfapps/ipdbproject/IEDIndex3.cfm?tid=5&pid=53&aid=1. Accessed August 11, 2012.

1.13 "World Population Prospects: The 2008 Revision." *Population Newsletter* 87. Population Division, Department of Economic and Social Affairs, United Nations Secretariat. http://www.un.org/esa/population/publications/popnews/Newsltr_87.pdf. Accessed December 19, 2012.

1.14 "World Energy Outlook." International Energy Agency (IEA), http://www.iea.org/publications/worldenergyoutlook/resources/energydevelopment/. Accessed December 19, 2012.

1.15 "Worldwide Trends in Energy Use And Efficiency." International Energy Agency (IEA). http://www.iea.org/publications/freepublications/publication/Indicators_2008.pdf. Accessed December 19, 2012.

1.16 "Energy Efficiency: A Recipe for Success." World Energy council. http://www.worldenergy.org/documents/fdeneff_v2.pdf. Accessed December 19, 2012.

1.17 "Frequently Asked Questions," U.S. Energy Information Administration, http://www.eia.gov/tools/faqs/faq.cfm?id=447&t=1. Accessed May 22, 2013.

1.18 "World Energy Book: Unconventional Hydrocarbons: A Hidden Opportunity." http://www.petroleum-economist.com/Article/2745687/World-Energy-Book-Unconventional-hydrocarbons-a-hidden-opportunity.html. Accessed December 19, 2012.

1.19 "Unconventional Resources." CGGVeritas. http://www.cggveritas.com/default.aspx?cid=3527&lang=1. Accessed December 19, 2012.

1.20 "2010 Survey of Energy Resources." World Energy Council. http://www.worldenergy.org/documents/ser_2010_report_1.pdf. Accessed January 8, 2013.

1.21 Maugeri, L. 2012. "Oil: The Next Revolution. The Unprecedented Surge of Oil Production and What It Means to the World." The Geopolitics of Energy Project. Kennedy School of Government, Harvard University,

Cambridge, MA. http://belfercenter.ksg.harvard.edu/files/Oil-%20The%20Next%20Revolution.pdf, Accessed December 19, 2012.

1.22 Folger, P. 2010. "Gas Hydrates: Resource and Hazard." Congressional Research Service. http://www.fas.org/sgp/crs/misc/RS22990.pdf. Accessed December 19, 2012.

1.23 Sandrea, I., and Sandrea, R. 2007. "Global Oil Reserves—Recovery Factors Leave Vast Target for EOR Technologies." *Oil and Gas Journal*, Part 1: November 5, 2007; Part 2: November 12, 2007. http://www.ipc66.com/publications/Global_Oil__EOR_Challenge.pdf. Accessed December 19, 2012.

1.24 "Coal: Energy for Sustainable Development (2012)." World Coal Association. http://www.worldcoal.org/resources/wca-publications/. Accessed January 8, 2013.

1.25 "The Global Status of CCS: 2011." Global CCS Institute. http://www.globalccsinstitute.com/publications/global-status-ccs-2011. Accessed December 19, 2012.

1.26 Smil, V. 2011. "Global Energy: The Latest Infatuations." *American Scientist* 99: 212. http://www.vaclavsmil.com/wp-content/uploads/docs/smil-article-2011-AMSCI.11.pdf. Accessed December 19, 2012.

1.27 "Electric Generation Efficiency." National Petroleum Council (NPC). http://www.npc.org/study_topic_papers/4-dtg-electricefficiency.pdf. Accessed December 19, 2012.

1.28 "Oil Shales/Tar Sands Guide." U.S. Department of the Interior. http://ostseis.anl.gov/guide/index.cfm. Accessed December 19, 2012.

1.29 Ridley, M. 2011. "The Shale Gas Shock." GWPF. http://thegwpf.org/images/stories/gwpf-reports/Shale-Gas_4_May_11.pdf. Accessed December 19, 2012.

1.30 "World Shale Gas Resources: An Initial Assessment of 14 Regions Outside the United States." U.S. EIA, 2011. http://www.eia.gov/analysis/studies/worldshalegas/. Accessed December 19, 2012.

1.31 "Shale Gas." Wikipedia, http://en.wikipedia.org/wiki/Shale_gas. Accessed December 19, 2012.

1.32 "FAQs: Natural Gas." International Energy Agency. http://www.iea.org/aboutus/faqs/gas/. Accessed December 19, 2012.

1.33 "Coal-Bed Methane: Potential and Concerns." US Geological Survey. http://pubs.usgs.gov/fs/fs123-00/fs123-00.pdf. Accessed December 19, 2012.

1.34 "Future Supply and Emerging Resources: Coal Bed Natural Gas." U.S. DOE. http://www.netl.doe.gov/technologies/oil-gas/futuresupply/coalbedng/coalbed_ng.html. Accessed July 16, 2012.

1.35 "What Is Coal Seam Gas?" Australia Pacific LNG. http://www.aplng.com.au/home/what-coal-seam-gas. Accessed December 19, 2012.

1.36 "Unconventional Gas." National Petroleum Council, 2007. http://www.npc.org/study_topic_papers/29-ttg-unconventional-gas.pdf. Accessed December 19, 2012.

1.37 "What Is Tight Gas and How Is It Produced?" Rigzone. http://www.rigzone.com/training/insight.asp?insight_id=346&c_id=4. Accessed December 19, 2012.

1.38 "Annual Energy Outlook 2012." DOE/EIA-0383(2012), June 2012. http://www.eia.gov/forecasts/aeo/pdf/0383(2012).pdf. Accessed on December 19, 2012.

1.39 "Levelized Cost of New Generation Resources in the Annual Energy Outlook 2011." U.S. EIA. http://www.eia.gov/oiaf/aeo/electricity_generation.html. Accessed on December 19, 2012.

1.40 "Energy Efficiency Indicators for Public Electricity Generation from Fossil Fuels." International Energy Agency (IEA), 2008. http://www.iea.org/publications/freepublications/publication/En_Efficiency_Indicators.pdf. Accessed on January 9, 2013.

1.41 "Energy Statistics Manual." International Energy Agency (IEA). http://www.iea.org/stats/docs/statistics_manual.pdf. Accessed on December 19, 2012.

1.42 "Oil Market Report—13 June 2012." International Energy Agency (IEA). http://omrpublic.iea.org/currentissues/full.pdf. Accessed on August 10, 2012.

1.43 "Reducing CO_2 Emissions from Passenger Cars." European Commission. http://ec.europa.eu/clima/policies/transport/vehicles/cars/index_en.htm. Accessed January 8, 2013.

1.44 "Worldwide Emissions Standards." Delphi. http://delphi.com/manufacturers/auto/powertrain/emissions_standards/. Accessed January 8, 2013.

1.45 "Biofuels Information." European Biofuels Technology Platform. http://www.biofuelstp.eu/global_overview.html. Accessed January 8, 2013.

1.46 "Biofuels for Transport: Policies and Possibilities." Organisation for Economic Co-operation and Development (OECD), 2007. http://www.oecd.org/dataoecd/18/8/39718027.pdf. Accessed January 8, 2013.

1.47 "Biofuel." Wikipedia. http://en.wikipedia.org/wiki/Biofuel. Accessed January 8, 2013.

1.48 "Technology Roadmap: Biofuels for Transport, 2011." International Energy Agency (IEA). http://www.iea.org/papers/2011/biofuels_roadmap.pdf. Accessed July 2012.

1.49 Giampietro, M., and Mayumi, K. 2009. *The Biofuel Delusion: The Fallacy of Large Scale Agro-Biofuels Production*. Earthscan, London and Sterling, VA.

1.50 Renewable Fuels Association. June 2012. http://www.ethanolrfa.org/news/entry/global-ethanol-production-to-reach-85.2-billion-litres-in-2012/. Accessed July 21, 2012.

1.51 Eisentraut, A. 2010. *Sustainable Production of Second Generation Biofuels—Potential and Perspectives in Major Economies and Developing Countries.* International Energy agency (IEA). Also http://www.oecd.org/berlin/44567743.pdf. Accessed January 9, 2013.

1.52 "Fischer-Tropsch Process." Wikipedia. http://en.wikipedia.org/wiki/Fischer%E2%80%93Tropsch_process. Accessed July 22, 2012.

1.53 "A Fine for Not Using a Biofuel That Doesn't Exist." *New York Times*, January 9, 2012. http://www.nytimes.com/2012/01/10/business/energy-environment/companies-face-fines-for-not-using-unavailable-biofuel.html. Accessed January 9, 2013.

1.54 "CIA: The World Fact Book." Central Intelligence Agency. https://www.cia.gov/library/publications/the-world-factbook/geos/us.html. Accessed January 9, 2013.

1.55 "Algal Biofuels." U.S. Department of energy. http://www1.eere.energy.gov/biomass/pdfs/algalbiofuels.pdf. Accessed January 8, 2013.

1.56 Hannon, M., Gimpel, J., Tran, M., Rasala, B., and Mayfield, S. 2010. "Biofuels from Algae: Challenges and Potential." *Biofuels* 1(5):763–784. http://www.ncbi.nlm.nih.gov/pmc/articles/PMC3152439/?tool=pubmed. Accessed January 5, 2013.

1.57 "Sustainable Development of Algal Biofuels in the United States." National Academies Press, 2012. http://www.nap.edu/catalog.php?record_id=13437. Accessed January 9, 2013.

1.58 "Gas-to-Liquids." Shell Global. http://www.shell.com/home/content/future_energy/meeting_demand/natural_gas/gtl/. Accessed July 23, 2012.

1.59 "Coal to Liquids." World Coal Association. http://www.worldcoal.org/coal/uses-of-coal/coal-to-liquids/. Accessed January 8, 2013.

1.60 "About Synthetic Fuels: Availability." Alliance for Synthetic Fuels Europe (ASFE). http://www.synthetic-fuels.eu/about_synthetic_fuels/availability_en.php. Accessed July 23, 2012.

1.61 "Gas to Liquids (GTL)." National Petroleum Council. 2007. http://www.npc.org/Study_Topic_Papers/9-STG-Gas-to-Liquids-GTL.pdf. Accessed January 8, 2013.

1.62 Economides, M.J. 2005. "The Economics of Gas to Liquids Compared to Natural Gas." World Energy Magazine 8(1):136–140. http://www.worldenergysource.com/articles/pdf/economides_we_v8n1.pdf. Accessed June 6, 2013.

1.63 "About NGVs." Natural Gas Vehicles for America. http://www.ngvc.org/about_ngv/index.html. Accessed July 24, 2012.

1.64 "Liquefied Petroleum Gas (LPG) in Transport." ClimateTechWiki. http://climatetechwiki.org/technology/lpg. Accessed July 24, 2012.

1.65 "Well-to-Wheels Analysis of Future Automotive Fuels and Powertrains in the European Context." CONCAWE and EUCAR, European Commission 2007. http://ies.jrc.ec.europa.eu/uploads/media/WTW_Report_010307.pdf. Accessed January 8, 2013.

1.66 "State of Charge." Union of Concerned Scientists, 2012. http://www.ucsusa.org/assets/documents/clean_vehicles/electric-car-global-warming-emissions-report.pdf. Accessed January 8, 2013.

1.67 "How to Avoid an Electric Shock. Electric Cars from Hype To Reality." European Federation for Transport and Environment, 2009. http://www.transportenvironment.org/sites/default/files/media/2009%2011%20Electric%20Shock%20Electric%20Cars.pdf. Accessed January 8, 2013.

1.68 Palou-Rivera, I., and Wang, M. 2010. "Updated Estimation of Energy Efficiencies of U.S. Petroleum Refineries." http://www.transportation.anl.gov/pdfs/TA/635.PDF. Accessed January 8, 2013.

Chapter 2
Manufacture, Composition, and Properties of Practical Fuels for Internal Combustion Engines

The fuels of choice for internal combustion engines are liquid hydrocarbons, because of their high volumetric energy density and ease of handling, storage, and transport. As a result, a worldwide supply and distribution network has built up over the last century for liquid fuels. The first oil well was dug in the United States in 1859 [2.1], and oil was distilled to make kerosene for lighting. Gasoline, which is much lighter than kerosene, was only recognized as a valuable fuel with the advent of the automobile in the last decade of the 19th century. Rudolf Diesel developed the diesel engine in the 1890s and originally intended his engine to run on all sorts of fuels ranging from kerosene to coal dust. He was particularly enthusiastic about running it on vegetable oils, believing that farmers would benefit from making their own fuel [2.2].

By 1920 there were 9 million cars in the United States alone [2.1], and the petroleum refining industry had to expand to supply the increasing volume, as well as improve the quality of the fuel required by the internal combustion engine as it developed and spread. In spark ignition (SI) engines, knock (see Chapter 4) became a problem as engine designers sought higher efficiency. Initially, the antiknock quality of the fuels was improved by adding aromatics produced from the distillation of coal tar [2.3]; later, tetraethyl lead (TEL) was discovered by Thomas Midgeley of General Motors in 1921. The use of lead additives to improve the antiknock quality of gasoline became common in spite of initial widespread opposition on health grounds [2.4].

Knock became of sufficient concern for antiknock quality to become the most important property of gasoline, and the octane scale and the octane numbers were introduced in 1929. Fuel antiknock quality improved continually as refining processes were improved, which in turn led to increased efficiency of the SI engine. Eventually, three-way catalysts had to be used to reduce pollutants from modern SI engines. However, lead and sulfur interfere with these catalysts, and the spread of such aftertreatment technology from the 1970s onward, combined with renewed concerns about toxicity, led to the removal of lead and a great reduction in sulfur levels in gasoline in most major gasoline markets. An evolution in fuels for diesel engines has also taken place regarding sulfur levels, viscosity, deposit-forming tendencies, low-temperature performance, lubricity, and ignition quality as the diesel engine developed over the past century. Gasoline and diesel fuels are complex mixtures of hydrocarbons. Table 2.1 lists compounds found at greater than 1% by weight in a European super unleaded gasoline.

These compounds make up just over 70% of the fuel by weight; the rest is made up by more than 100 other compounds that can be identified. The main classes of hydrocarbons that can be found are paraffins (alkanes); olefins (alkenes); naphthenes (cycloalkanes); and aromatics, which could be further split into mono and polynuclear aromatics and oxygenates, that is, alcohols and ethers in gasoline and esters in diesel fuel. Fuels also contain sulfur and nitrogen compounds and trace levels of metal compounds. It is common for fuels to contain additives, which are chemicals added in small quantities (<0.5% by volume), either to improve performance or correct a deficiency.

Table 2.2 lists some other properties and composition of such fuels that are usually measured. Gasoline and diesel differ from each other primarily in terms of their autoignition quality, which is measured by the research octane number (RON) and motor octane number (MON) for gasolines and by the cetane number (CN) for diesel. Gasoline needs to be resistant to autoignition while diesel is required to autoignite easily. Gasoline is much more volatile than diesel, as illustrated in Fig. 2.1, which shows the distillation characteristics of typical European fuels.

Fuels sold in a given market would be required to be within limits set by fuel specifications for many of these properties. In many markets fuels are now required to contain bio-components, usually oxygenates, which are manufactured outside the conventional refinery (for more details, see Chapter 1). We will now examine the manufacture, properties, and effects of those properties on engine performance and emissions.

Table 2.1 Compounds Present at Greater Than 1% by Weight in a European Super Unleaded (~98 RON) Gasoline

Compound	weight %
toluene	11.7
isopentane	11.43
m-xylene	6.93
n-butane	5.41
iso-octane	4.34
o-xylene	4.04
ethylbenzene	3.04
1,2,4-trimethylbenzene	2.69
p-xylene	2.58
2-methylbut-2-ene	2.3
2-methylpentane	2.07
n-pentane	2.02
isobutane	1.87
trans-pent-2-ene	1.7
m-ethyltoluene	1.65
2-methyl-1-ene	1.39
2-methylbut-1-ene	1.39
3-methylpentane	1.31
2,3,4-trimethylpentane	1.3
2-methylhexane	1.24
benzene	1

2.1 Manufacture and Composition of Practical Fuels

Liquid fuels for transport are primarily made in refineries, starting with crude oil [2.5] and blending the products of several refinery processes [2.3, 2.6]. Refineries also supply feed stocks for other industries and make specialty chemical products. On a worldwide basis, around 95% of transport energy comes from petroleum, with this proportion projected to decrease to 90% by 2040 [2.5].

Crude oil itself contains thousands of different hydrocarbons, each with its own boiling point. Initially, the crude oil is separated into different boiling range fractions by distillation: either atmospheric distillation or vacuum

Table 2.2 Some Properties of a Typical Gasoline and Diesel in Europe

Properties	Gasoline	Diesel
RON	95.4	
MON	85.6	
cetane number (CN)		54
energy content (to gas) MJ/kg	43.5	43.2
density @ 15 °C, kg/liter	0.738	0.833
initial boiling point (IBP), °C	28	165
final boiling point (FBP), °C	198	352
normal paraffins, %v	10.8	
isoparaffins, % v	43.4	
total paraffins, % v	54.2	44
naphthenes, % v	2.9	29
olefins, % v	8.6	
Aromatics, %v total	33.6	25.6
monoaromatics		22.1
diaromtics		3.2
triaromatics		0.3
Benzene, % v	0.88	
Sulfur	20	9
Average molecular formula		
C	6.64	15.4
H	12.11	30.1

distillation at pressures between 7 and 13 kPa and the temperature limited to 350 °C [2.6]. Up to 2% of the crude may be dissolved gases that are released when the temperature is raised above ambient temperature; these constitute liquefied petroleum gas (LPG). The main component (~75%) of LPG is propane. The light fraction resulting from distillation and boiling roughly in the gasoline range from about 20–160 °C is often termed *naphtha*. The middle distillates span the diesel boiling range of about 160–380 °C and form the kerosene/gas oil range. Between 40% and 60% of the weight of the crude oil may be residue with boiling points above about 380 °C, depending on the crude.

The demand for liquid transport fuels cannot be met without further processing of these streams to convert heavier, low-value fractions into lighter fractions that can be used as fuel components. Most of the complex refinery

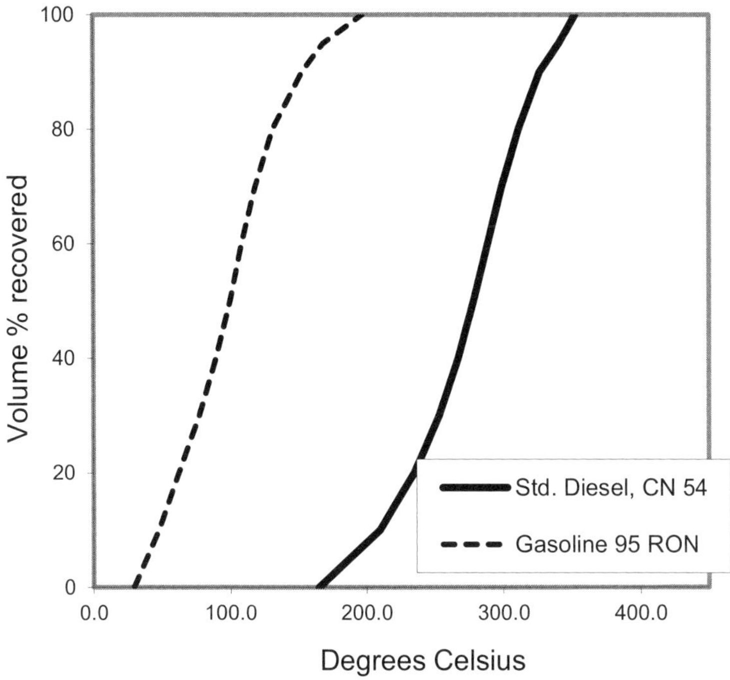

Fig. 2.1. Volatility characteristics of a typical gasoline and a typical diesel fuel in Europe.

processing is aimed at such conversion and toward increasing the antiknock quality of the streams from the initial distillation. Antiknock quality is related to the molecular structure of the fuel [2.7, p. 470]. For instance, branched paraffins have much better antiknock quality than straight-chain paraffins with the same number of carbon atoms; increasing the length of the carbon chain increases the knocking tendency; lengthening the side chain attached to the ring structure of both naphthenes and aromatics increases the knocking tendency. Refinery processes involve separation into smaller fractions, a change of composition, removal of contaminants, and a redistribution of carbon/hydrogen ratios of the molecules through carbon rejection or hydrogen addition [2.6]. A description of refinery and other processes used for making fuel components can be found in [2.3, 2.6], but a brief summary of the main processes is given here next.

Coking processes are a series of thermal processes used to convert nonvolatile residues into fluid distillable products. These processes include delayed coking, fluidized solids coking, and flexicoking. In these processes, the residue is heated in either a "coke drum" or a fluidized bed, and the vapors are collected. Petroleum coke, which is high in carbon, is left behind; it is used as fuel or in the manufacture of components such as electrodes.

Visbreaking reduces the viscosity of the residual oil by thermally cracking large hydrocarbon molecules by heating the residual oil to 450–510 °C at pressures of 3–20 bar (0.3–2 MPa). It does not use a catalyst.

Catalytic cracking involves a heavy oil that is pyrolyzed on a catalyst at 700–1000 K to produce a range of lighter products with high proportions of aromatics and olefins (alkenes). Fluid catalytic cracking has supplanted the thermal cracking processes that were used to convert distillate oils in early refineries.

Catalytic reforming is widely used to increase the antiknock properties of naphtha. The chemical processes involved include the conversion of naphthenes (cycloparaffins) into aromatics by dehydrogenation, isomerization of straight chain hydrocarbons into branched hydrocarbons, and removal of side chains from aromatics. The process produces much of the hydrogen used elsewhere in the refinery.

Hydroprocessing entails heating high-boiling-point feedstock with hydrogen in the presence of a catalyst. The presence of hydrogen terminates many of the coke-forming reactions and yields lower-boiling-point components. Hydrocracking employs temperatures greater than 350 °C and pressures of between 70 and 200 bar (7–20 MPa). Some of the product from the hydrocracker is often put through a further catalytic reforming process. Hydrotreating is specifically aimed at converting sulfur and nitrogen compounds to hydrogen sulfide and ammonia, which can be removed, leaving components with lower sulfur and nitrogen. The feed is heated to around 350 °C at 35–70 bar (3.5–7 MPa) in the presence of hydrogen, and selectivity is controlled by the catalyst and the temperature.

Alkylation converts light olefins (alkenes) derived from catalytic cracking to C6–C9 branched paraffins (alkanes).

Isomerization converts straight-chain paraffins to branched-chain paraffins with the help of a catalyst.

These different processing units are interconnected in a fully complex refinery, as illustrated in Fig. 2.2.

Such an arrangement leads to the conversion of low-value streams that result from simple distillation into high-value products such as gasoline and diesel. Table 2.3 shows typical product yields for a fully complex refinery for a typical North Sea crude oil.

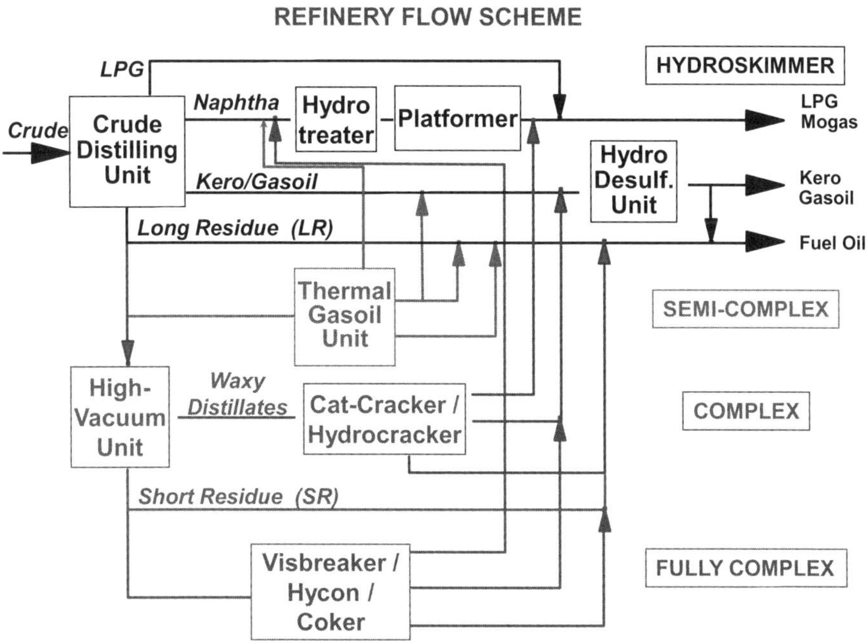

Fig. 2.2. The arrangement of a fully complex refinery.

Depending on whether there is a catalytic cracker unit (CCU) or a hydrocracker unit (HCU) aligned with a visbreaker unit (VBU), the yields will be different. However, compared to simple distillation, the yields of useful products like gasoline are significantly increased. Refinery yields also depend very much on the crude. For example, the gasoline yield could vary between 25% and 35% in the same refinery depending on the crude oil used.

Table 2.3 Yields from a Complex Refinery—Illustration for a North Sea Crude

Yield (% m)		Simple Distillation		Fully Complex HCU / VBU		Fully Complex CCU / VBU
LPG		1.5		2.3		5
naphtha		23.2		6.3		7.2
gasoline		0		20.4		28.8
kero/jet		11		16.7		11.1
diesel		24.8		40		27.5
fuel oil		37.5		6		13.7
fuel and loss		2		8.3		6.7

Modern refineries need to be flexible enough to process different crudes to meet the demand for different products and ensure that these products meet required specifications.

The boiling ranges of different refinery products are shown in Fig. 2.3.

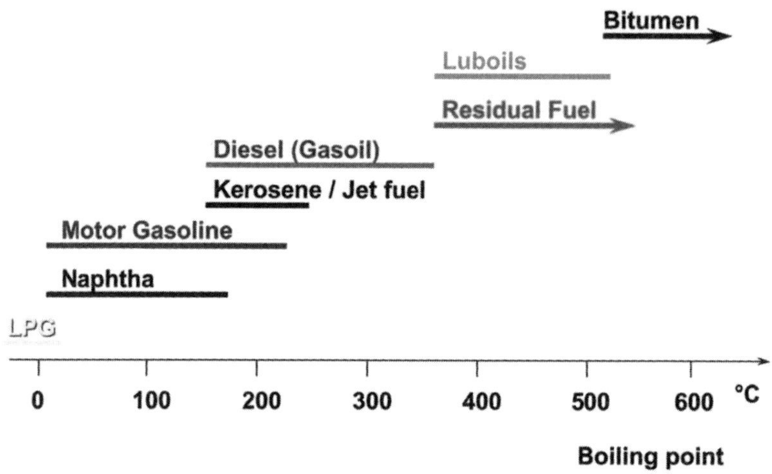

Fig. 2.3. Boiling ranges of different refinery components.

In addition, other processes outside a conventional refinery produce fuel components that are often used in liquid fuels.

Ether synthesis produces branched ethers like methyl tertiary butyl ether (MTBE) by reacting alcohols with branched olefins such as isobutene. MTBE is a very useful component for increasing the antiknock quality of gasoline and is still used in many parts of the world. However, other parts of the world, such as the United States, have stopped using MTBE because of concerns about the contamination of the water table.

Gas-to-liquid (GTL) method makes use of the Fischer-Tropsch process for synthesizing syngas, a mixture of carbon monoxide and hydrogen into liquid fuels that dates back to the 1920s. It can produce gasoline components or diesel components depending on the catalyst and catalyst temperature used (for more details, see Chapter 1 and [2.8]).

Biofuels are made from sugar, starch, vegetable oils, cellulosic residue, and biological waste, and are increasingly being used as fuel components.

2.2 Fuel Composition and Properties and Their Effects on Engine Performance and Emissions

Fuel composition and some fuel properties are known to affect engine performance and emissions [2.3]. The criteria for measuring these properties have been standardized, for example, by ASTM International (formerly, the American Society for Testing and Materials; see [2.9]) or the Energy Institute with its International Petroleum (IP) test methods [2.10]. In most cases these methods have been harmonized so that an ASTM test would have an equivalent IP test. Fuel quality is defined by selected properties as well as by composition, and in every part of the world fuel specifications try to ensure that there is a minimum fuel quality available. Such standards require that a given fuel property, measured according to a standard test procedure, must fall within a specified range for the fuel to be legally sold in the particular market. Such specifications are intended to ensure, as far as possible, that an engine will start easily when cold, warm up rapidly, run smoothly, deliver adequate power under all conditions, provide good fuel economy, and generate low emissions. Specifications also stipulate that the fuel will not form harmful engine deposits, cause excessive wear, harm aftertreatment systems intended to control engine emissions, or contaminate or corrode the fuel system.

The European Committee for Standardization (CEN; [2.11]) sets such standards in Europe, where three standards cover the quality of automotive fuels: EN 228 for gasoline, EN 590 for diesel fuel, and EN 589 for LPG. In the United States, the standard specification is set by ASTM International: The fuels for SI engines are covered by ASTM D4814, and diesel fuel by ASTM D975 [2.9]. In addition to these standards, fuels in the United States have to comply with the rules, regulations, and waivers set by the U.S. Environmental Protection Agency (EPA; see [2.12]) and comply with pertinent state or local regulations, such as those set by the California Air Resources Board (CARB).

The Air Quality Improvement Research Program (AQIRP) in the United States, also known as the Auto/Oil Program [2.13], and the European Programme on Emissions, Fuels and Engine Technologies (EPEFE) [2.14] were pioneering large-scale systematic studies on the effects of fuel properties on engine emissions. AQIRP was a joint study among 14 oil companies and the three major U.S. auto manufacturers that ran from 1989 to 1992. It mostly considered U.S. vehicles made between 1983 and 1989. The European program followed later and hence had a more modern fleet of vehicles. A similar program was later initiated in Japan in 1996 [2.15]. Fuel standards in the United States, Europe, and Japan have been greatly influenced by these studies. There is a wide variation in fuel quality across the world, and attempts have been made by engine and equipment manufacturers to harmonize fuel quality [2.16].

We consider in the next two sections some of the important properties and compositional factors for gasoline and diesel fuel. However, remember that many of the properties and compositional parameters are highly correlated. When conducting studies of the fuel effects on engine performance and emissions, it is difficult to vary any parameter completely independently to isolate its effects. Moreover, combustion, performance, and emissions depend on engine design, operating conditions, and control strategies, as well as on the fuel—and fuel effects can be different in different engines. An excellent review of the literature on the fuel effects on emissions can be found in [2.17]. Fuel specifications are occasionally mentioned in the following discussion to give an idea of typical ranges involved. Detailed fuel specifications in any geographical area can usually be found on the Internet.

2.2.1 Gasoline

2.2.1.1 Gasoline Antiknock Properties

Knock is an abnormal combustion phenomenon that can damage SI engines [2.7]. Engine efficiency is reduced because of the requirement to avoid knock, which is initiated by autoignition in the unburned mixture known as the *end-gas*, ahead of the flame front inside the engine cylinder. Knock is determined by the pressure and temperature history of the end-gas and the antiknock or autoignition quality, which is traditionally measured by the RON and MON of the fuel used. The RON test is run in a single-cylinder Cooperative Fuel Research (CFR) engine at an engine speed of 600 rpm and an intake temperature of 52 °C, and the MON test is run at 900 rpm and with a higher intake temperature of 149 °C. These tests are run in strict accordance with the procedures set in ASTM D2699 for RON and ASTM D2770 for MON [2.9]. The octane scale is based on two paraffins, n-heptane and iso-octane. Blends of these two primary components are referred to as primary reference fuels (PRFs) and define the intermediate points in the RON or MON scale. The RON or MON is the volume percent of iso-octane in the PRF. Thus a blend of 95% iso-octane and 5% n-heptane by volume is assigned the octane number of 95 in both RON and MON scales. A fuel is assigned the RON (or MON) value of the PRF that matches its knock behavior in the RON (or MON) test. RON and MON are among the most important properties of a practical gasoline since the manufacture and marketing of transport fuels is profoundly influenced by the requirements to meet RON and MON specifications.

All practical gasolines are mixtures of aromatics, olefins, naphthenes, oxygenates, and paraffins. The autoignition chemistry of such practical fuels is different from that of PRF. A practical gasoline will match PRF of a higher octane number in the RON test and hence has a higher RON when compared to the MON test. The difference between RON and MON is known as the sensitivity. Practical fuels for SI engines have RON ranging from 90 to 100 with sensitivity in the range of 7 to 12; the most common fuels in Europe have RON ~ 95 and MON ~ 85. The antiknock quality of a fuel is best described by an octane Index, OI = RON – KS, where S is the sensitivity and K is an empirical constant that depends only on the engine design and operating conditions. In most modern engines, K is negative, so that, for a given RON, a lower MON, higher sensitivity fuel has better antiknock quality. These issues are discussed in greater detail in Chapter 4.

2.2.1.2 Volatility

The distillation (volatility) characteristics of fuels have an important effect on their safety and performance. The boiling range of fuels is determined by distillation at atmospheric pressure according to the procedure specified in ASTM D86 [2.9]; it is among the oldest test methods defined by the ASTM. The test yields the total volume percent of the fuel that is evaporated at a given temperature, as shown in Figs. 2.1 and 2.4. Two measures are used to define volatility: the T number, T*xx*, which is the temperature at which *xx*% volume of the fuel is evaporated, and the E number, E*yy*, which is the volume percent of the fuel that is evaporated at a temperature of *yy*. In the example shown in Fig. 2.4, T50 = 100 °C and E100 = 50%.

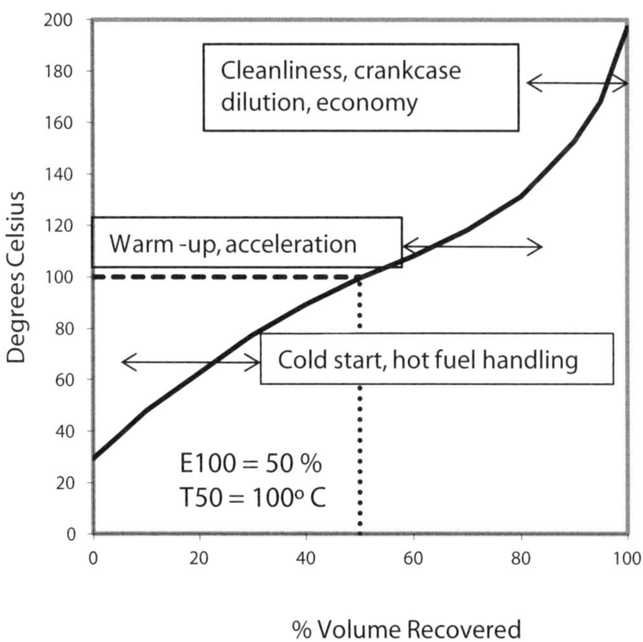

Fig. 2.4. Measures of volatility and effects of volatility on performance.

Another important parameter related to volatility is the Reid vapor pressure (RVP), which is the absolute vapor pressure exerted by the fuel at 100 °F (37.8 °C) and is measured according to ASTM D5191 [2.9].

In general, volatility at the lower temperatures affects cold start and hot fuel handling. If, for instance, RVP is too low or E50 is too small, there might be problems starting the engine at cold conditions, but if they are too large, there might be problems with vapor lock in the fuel system when the engine gets too hot. Higher RVP will also cause evaporative emissions to increase in older vehicles [2.18, 2.19]. However, U.S. vehicles, which were adapted to federal regulations in 1996, showed very little effect of RVP on evaporative emissions [2.19]. Specifications on RVP are usually different for summer and for winter to cater to the volatility needs of vehicles at different ambient temperatures. Addition of alcohols and other oxygenates such as MTBE increases the volatility at low temperatures [2.3]. Low midrange volatility can cause poor warm-up and acceleration. If the volatility is too low at high temperatures because the fuel contains high concentrations of high boiling point materials, there might be problems with oil dilution or excessive deposit formation (see Chapter 3). If, on the other hand, it is too high, fuel economy could suffer, but this might be because the density of the fuel will be low in such a case.

A driveability index (DI), which is a combination of different volatility parameters, is often used to consolidate volatility effects on vehicle operation. The driveability of the vehicle itself is defined by subjective ratings by trained raters who drive it through a specified test cycle. The raters give demerit points, which are aggregated, on hesitation (lack of response to opening the throttle), stumble (intermittent loss of power), surge (cyclic pulses of power), idle roughness, and difficulty of starting and engine stalls [2.20, 2.21]. Driveability demerits usually show a correlation with DI. The ASTM specification of DI is given by:

$$\mathrm{DI} = 1.5\mathrm{T}10 + 3.0\mathrm{T}50 + 1.0\mathrm{T}90 + 2.4(\text{ethanol vol\%}) \tag{2.1}$$

where the T numbers are in degrees Fahrenheit. The term with the ethanol concentration was included following a research program conducted on behalf of the Coordinating Research Council (CRC) in 2003 [2.22]. The maximum value set for DI in ASTM D4814 is 1250 [2.9]. In addition to these parameters, high values of T70 and T30 have also been found to contribute to a deterioration of driveability and driver dissatisfaction [2.23]. Poor cold start and warm-up driveability also results in increased hydrocarbon (HC) emissions [2.24, 2.25].

2.2.1.3 Density

Low density could lead to increased volumetric fuel consumption, and high density could lead to oil dilution and excessive combustion chamber deposits. The EN 228 specification in Europe requires gasoline density to be between 0.72 and 0.7750 g/cc.

2.2.1.4 Gum Specifications

Gum is a generic name for high-boiling-point products of oxidation that are soluble in gasoline. High gum levels could indicate high-deposit-forming potential in different parts of the engine (see Chapter 3). Hence, the maximum gum level is usually specified. Unwashed gum is measured by evaporating a sample of gasoline and weighing the residue according to the procedure set out in ASTM D381 [2.9]. However, detergent additives used for deposit control usually contain relatively nonvolatile components and can cause a significant increase in unwashed gum level. Unwashed gum is washed with a solvent to get washed gum. The EN 228 specifications and the Worldwide Fuels Charter [2.16] set a maximum limit of 5 mg/100 ml for washed gum.

2.2.1.5 Aromatic Content

In general aromatics, fuel molecules containing at least one benzene ring contribute significantly to fuel antiknock quality and have high volumetric energy content. Both the AQIRP and EPEFE tests showed that lowering aromatic levels reduces toxic benzene emissions in vehicle exhausts [2.17, 2.26, 2.27]. However, other air toxics like acetaldehyde, formaldehyde, and 1,3-butadiene were slightly increased when aromatic level was lowered [2.26], but the effects on benzene are dominant. Heavy aromatics can also increase combustion chamber deposits. In both United States and Europe, the maximum level of aromatics in gasoline is set at 35% by volume. Aromatics also produce more CO_2 for a given energy content because of their higher carbon-to-hydrogen ratios. In addition, benzene levels are controlled [2.28]; in Europe the maximum allowable is 1% by volume.

2.2.1.6 Olefin Content

Olefins are unsaturated hydrocarbons and are thermally unstable. They generally have good antiknock quality but can cause increased deposit formation, particularly in fuel injectors. Their combustion can cause increased toxic emissions [2.26], and they have high ozone-forming potential [2.29]. The maximum level in European gasolines is set at 18% by volume.

2.2.1.7 Sulfur Content

Sulfur reduces the efficiency of three-way catalysts and hence increases all emissions [2.17]. Sulfur is naturally present in crude oil and active steps have to be taken to reduce its concentration in the finished fuel. In the AQIRP study [2.30], the fleet average HC, CO, and NOx were reduced by 16, 13, and 9%, respectively, on reducing the sulfur level from 466 to 49 ppm. In the EPEFE study, when fuel sulfur level decreased from 382 to 18 ppm, fleet average emissions of CO, HC, and NOx were reduced by approximately 10% [2.31]. Other studies conducted on more modern fleets of cars, which conform to more stringent emissions standards, confirm this trend [2.32, 2.33, 2.34]. In fact, since the baseline is low, the percentage deterioration of emissions performance is much more marked in modern cars than in the older fleets considered in the AQIRP and EPEFE tests when sulfur is increased. Sulfur also adversely affects oxygen sensors [2.16]. Therefore, fuel sulfur levels have been significantly reduced in most developed markets in the past two decades [2.35]; see Table 2.4. The maximum level allowed in Europe by the EN 228 specification is 50 ppm.

Table 2.4 Fuel Sulfur Levels in Gasoline in the United States and Europe (Data from [2.35])

United States		Europe	
YEAR	Maximum Sulfur, ppm	YEAR	Maximum Sulfur, ppm
1975, First catalysts (oxidation)	1000	Pre 1993 (leaded)	1000–2000
1980, Tier 0, 3-way catalyst	1000	1993, Euro 1 (3-way catalyst)	1000
1994, Tier 1	1000	1996, Euro 2	500
1999	1000	2000, Euro 3	150
From 2004, Tier 2*	80 (30)*	2005, Euro 4	50

*The standard calls for an average of 30 ppm and a not-to-be-exceeded level of 80 ppm. This results in a sulfur level of less than 30 ppm in some of the gasoline sold if some has to be at 80 ppm.

2.2.1.8 Oxygenates

Oxygenates like MTBE and ethanol have excellent antiknock properties and are used extensively as gasoline components, though the use of MTBE has been stopped in some countries because of concerns about contamination of the water table. However, ethanol can be made from bio sources and is attractive for reasons of energy security. It could also have a lower greenhouse gas footprint under some circumstances. Ethanol made from sugarcane has been used extensively in Brazil since the late 1970s. It is also increasingly used in the United States, with the U.S. EPA allowing 15% by volume of ethanol to be used in gasoline (E15) from October 2010 [2.36]. A number of companies are now investigating technologies that could enable butanol to be produced at commercial quantities from bio sources [2.37].

Oxygenates have much lower volumetric energy content compared to gasoline—the lower heating value of ethanol is 21.1 MJ/liter compared to around 32 MJ/liter for gasoline. They also affect volatility and hence evaporative emissions and driveability, as discussed in Section 2.2.1.2. In modern cars with feedback control systems, adding MTBE and ethanol to gasoline reduces CO and HC but increases NOx [2.17, 2.18]. More recent studies with low-emission vehicles in California show similar trends [2.38]. This study [2.38] also found that benzene, acetaldehyde, formaldehyde, and 1,3-butadiene, all considered air toxics, increased on the addition of ethanol. A CARB study reports that a gasoline containing 10% ethanol by volume decreased toxic emissions by 2% and CO by 10% but increased NOx by 14%, total HC by 10% and ozone forming potential by 9%, relative to a gasoline containing 11% MTBE by volume [2.39].

In general, primarily driven by energy security concerns, the use of ethanol is increasing across the world.

2.2.1.9 Metal Content

Metal additives like TEL and, to a lesser extent, MMT (methylcyclopentadienyl manganese tricarbonyl) and an iron compound, ferrocene (dicyclopentadienyl iron), have been used as antiknock additives. In general, auto manufacturers oppose the use of metallic, ash-forming additives because they reduce the efficiency of three-way catalysts used for emission control in modern vehicles [2.16]. Lead also has known adverse effects on health, and lead additives have been banned in many developed markets. It is fair to say that engine manufacturers would like to ban the use of MMT [2.16, 2.40] because of its effects on catalyst efficiency, spark plugs, and oxygen sensors. However, the manufacturers of MMT have put up a spirited defense of their product [2.41]. The U.S. EPA assessed potential health risks of particulate manganese (Mn) emitted from the combustion of MMT in gasoline and, as of September 2011, still allowed its use in U.S. gasolines at a maximum level of $^1/_{32}$ g of Mn per gallon, which is around 11 ppm Mn [2.42]. Ferrocene, like MMT is also used in some markets.

2.2.2 Diesel Fuel

2.2.2.1 Autoignition Quality (Cetane)

The autoignition quality of diesel fuels is determined by the CN. It is measured for a given fuel by comparing its ignition characteristics with reference fuels in a CFR cetane engine using the standard ASTM D613 test method [2.9]. The reference fuels are made by blending normal cetane, n-hexadecane, which is defined to have a CN of 100, and heptamethyl nonane, a highly branched paraffin that is assigned a CN of 15. The CN of a reference fuel with x% volume normal cetane is given by the following equation:

$$CN = x + 0.15(100 - x) \tag{2.2}$$

The fuel being tested is assigned the CN of the reference fuel with the same ignition quality in the cetane test.

More recently, a laboratory test method has been developed to measure the autoignition quality of diesel fuels—ASTM D6890-03a [2.9]. This test is also referred to as the ignition quality test (IQT). In this test, a small specimen of test fuel is injected into a heated, temperature-controlled, constant-volume chamber that has been previously filled with compressed air. The charge pressure and temperature are 2.137 ± 0.007 MPa and 545 ± 30 °C, respectively. The ignition delay (ID) is measured as the time between the start of fuel injection and the start of significant heat release; an average of 32 firings is taken. The derived CN, or DCN, is related to ID, in milliseconds, by the following formula:

$$\text{DCN} = 3.547 + 83.99(\text{ID} - 1.512)^{-0.658} \tag{2.3}$$

The repeatability and reproducibility of IQT measurements are better than those obtained in the cetane engine test, and DCN is gradually replacing CN as the preferred measure of diesel fuel autoignition quality. In general, the higher the CN or DCN, the lower the ID and the easier it is to ignite the fuel. Practical diesel fuels have CNs ranging from 40 to 60. The minimum specification for CN is 40 in the United States and 51 in Europe.

The cetane index (CI) is an estimation of CN using the distillation properties and the density of the diesel fuel (see ASTM D976-06, [2.9]). It is convenient because it does not require any additional testing of the samples in a CFR engine or an IQT test. However, if the fuel contains oxygenates or a cetane improver additive, CI is not a good measure for diesel autoignition quality. Even with fuels where it is most applicable, it is not a substitute for the CN and should be used as "a supplementary tool to estimate cetane number when used with due regard for its limitations" [2.9]. In cases where the CN of the fuel has been properly established, CI is useful as a subsequent check on the CN of the fuel, provided its source and mode of manufacture remain unchanged.

There is an inverse correlation between the CN and the octane number. Conventional diesel fuels cannot be run in the RON and MON tests because of their low volatility. Mixtures of toluene and heptanes, called toluene reference fuels (TRFs), and iso-octane and heptanes (PRFs) have had their RON and CN measured in [2.43]. Figure 2.5 shows the measured RON plotted against measured CN.

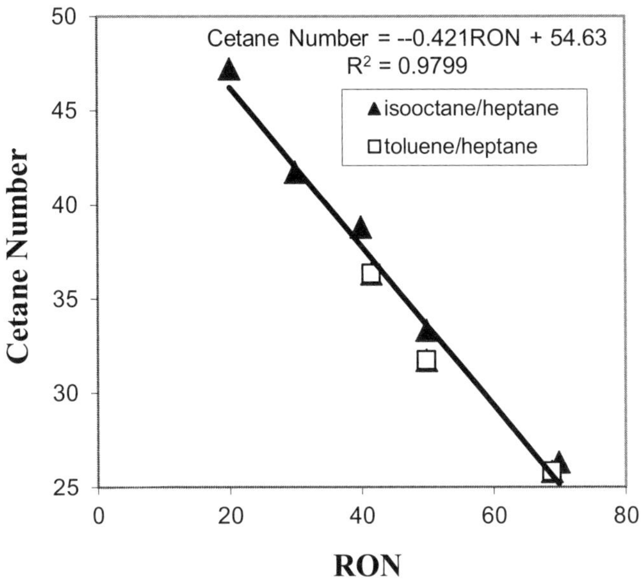

Fig. 2.5. Cetane number versus RON for PRF and TRF.

From Fig. 2.5,

$$CN = 54.6 - 0.42RON \tag{2.4}$$

Similar inverse correlations between RON and CN, over a wider range of CN, are given in [2.44]. Practical fuels with RON < ~ 60 and CN > ~ 30 can be considered diesel-like fuels [2.43], and their autoignition quality should really be specified by CN or DCN. In any case, their RON and MON cannot usually be determined because those tests cannot be run because of the low volatility of the fuel. The ID from the IQT test is plotted against DCN and RON in Fig. 2.6 using Eqs. 2.3 and 2.4. The variation in ID with CN is much less steep, in the range of practical diesel fuels with CN between 40 and 60, than in the range of practical gasolines with RON between 90 and 100.

The effects of autoignition characteristics of the fuel on combustion, emissions, and performance of compression ignition engines are explored in detail in Chapter 6. In summary, in conventional diesel engines, increasing the CN above 40 makes cold start easier; reduces engine noise at low load at cold conditions; and reduces NOx, HC, and CO [2.16]; but it most likely will increase soot.

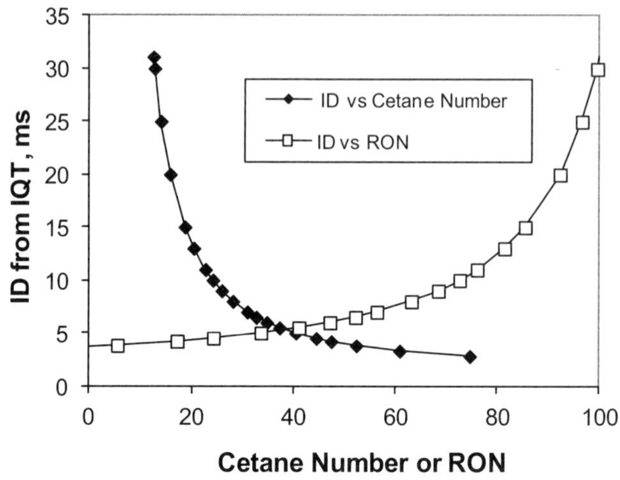

Fig. 2.6. Ignition delay from IQT test versus CN or RON.

2.2.2.2 Density and Viscosity

Variations in fuel density and viscosity will affect the mass of fuel injected because fuel injection is usually controlled by injection pulse width and, hence, volumetrically. In modern engines, this will affect power, fuel consumption, and emissions. Therefore density is controlled within a fairly narrow range. In the European specification (EN 590), the density at 15 °C has to be between 0.82 and 0.845 g/cc, whereas the kinematic viscosity at 40 °C is constrained to be between 2 and 4.5 mm^2/s. Reducing density appears to reduce HC, CO, and particulate emissions [2.17], probably because when density is reduced, heavy aromatic content is also reduced.

2.2.2.3 Distillation Properties

The EPEFE program [2.45] studied the effect of varying T95 between 320 and 375 °C. In the Euro 2 engines that were tested, increase in T95 over this range was associated with a 7% increase in particulates and around 5% decrease in NOx in light-duty engines. Emissions from heavy-duty engines were not significantly affected. The European specification (EN 590) puts a maximum value of 360 °C on T95.

2.2.2.4 Lubricity

High-pressure fuel pumps in diesel engines are lubricated by a complex mixture of polar compounds naturally found in the fuel. These occur at very low concentrations and form a protective layer on metal surfaces. Hydrotreating to reduce sulfur levels also removes many of these naturally occurring lubricants [2.46]. In fact, when very low-sulfur diesel fuel was first introduced in Sweden, there was a big problem in the field associated with rapid failure of injection pumps, which was eventually solved using appropriate additives (see Section 2.3). Diesel fuel needs to have sufficient lubricity to ensure the integrity of fuel pumps. Lubricity is measured by the size of the wear scar in the HFRR (high-frequency reciprocating rig) test, in which a hardened steel ball vibrates in loaded contact with a hardened steel plate immersed in the test fuel; the test method is described in ASTM D6709-11 [2.9] and IP 450. The European specification (EN 590) requires the wear scar in this test to be a maximum of 460 microns.

2.2.2.5 Cold Flow Properties

Up to a third of diesel fuel can be made up of normal paraffins with chain lengths typically ranging from 10 to 30 carbon atoms (C10–C30). These have limited solubility in the fuel, and if the fuel is cooled sufficiently, the heavier paraffins (C15 and higher) will come out of solution as wax, which could block fuel lines and filters. This could cause the vehicle to stall or malfunction at low temperatures [2.3, 2.4, 2.47, 2.48]. The cold flow performance of a diesel fuel depends on its composition and is described by different critical temperatures during the formation and growth of wax crystals as the fuel is cooled according to agreed test procedures. It can be improved through the use of additives.

The cloud point (CP) is the temperature at which the wax first becomes visible as the fuel is cooled in the ASTM D2500 (or IP 219) test procedure. It is a subjective measure in that it relies on the human eye and the opinion of the test operator. The repeatability of the CP determination—when the same sample is tested repeatedly by the same operator using the same apparatus—is 2 °C, and the reproducibility (between different operators in different laboratories) is 4 °C [2.3]. The pour point (PP) is the temperature at which, upon further cooling, the wax crystals grow and eventually interconnect to form a large network (gel) and stop the fuel from flowing; this is measured in the ASTM D97 test. In practice, neither CP nor PP is found to predict the actual cold weather performance of the fuel in a vehicle very well.

A more widely accepted cold flow property is the cold filter plugging point (CFPP), determined according to the ASTM D6379 (IP 309) test procedure. It measures the lowest temperature at which 20 ml of fuel will pass through a fine wire mesh screen of 45 micrometers nominal aperture in less than 60 seconds. Checks are made at 1 °C intervals as the fuel is cooled. The repeatability of the test is 1 °C, and the reproducibility is between 3 and 6 °C. CFPP usually lies between CP and PP. In the United States a modified test procedure (ASTM D4539), the low-temperature flow test (LTFT), is used to determine the filter plugging point. It is a more severe test than the CFPP because the cooling rate is slower and was found to be a better predictor of cold operability of U.S. diesel vehicles [2.48, 2.49]. In Europe (EN 590), the specification for CFPP is a maximum of 0 °C in the summer and –15 °C in the winter.

2.2.2.6 Flash Point

Flash point is a measure of the tendency of a fuel to form a flammable mixture with air under a fixed condition as in the ASTM D93 (IP 34) procedure. It is the lowest temperature at which the fuel can vaporize to form an ignitable mixture in air, and it is used to quantify the flammability hazard of the fuel. The European (EN 590) specification requires the flash point to be at least 55 °C for diesel fuel. This means that if the temperature is below 55 °C, as it would be in Europe even in summer, the fuel/air mixture above the liquid, say, in a vehicle fuel tank would be too lean to be flammable, and there would not be a fire hazard. The flash point for gasoline is usually very low, around −45 °C, so that at normal temperatures, the vapor in a fuel tank is too rich to be flammable, and thus it is safe in the presence of an accidental ignition source. Adding a more volatile liquid like gasoline or, even more dramatically, ethanol to diesel fuel can depress the flash point and dramatically increase the flammability hazard. Much larger quantities (e.g., 15% volume) of a less volatile fuel like diesel can be added to gasoline without changing its flash point significantly.

2.2.2.7 Sulfur

Sulfur levels in diesel fuel have been significantly reduced in North America, Europe, and Japan over the past few decades [2.35] (e.g., see Table 2.5). Fuel sulfur increases soot levels [2.31, 2.34, 2.45], and at high concentrations, above 0.3%, it significantly reduces engine life [2.16]. It also has a deleterious effect on advanced aftertreatment systems used for NOx and particulate reduction [2.16]. The oxidation of fuel sulfur produces SO_2, which can be further oxidized to SO_4, which in turn combines with water to coalesce around the carbon core of the particulate. Oxidation catalysts significantly increase the conversion of

Table 2.5 Fuel Sulfur Levels in Diesel Fuel in the United States and Europe (Data from [2.35])

United States Heavy Duty		Europe	
YEAR	Maximum Sulfur Level, ppm	YEAR	Maximum Sulfur Level, ppm
		1980	5000
1988	5000	1989	3000
1993	500	1993, Euro 1	2000
1998	500	1996, Euro 2	500
2004/2006	500/15	2000, Euro 3	350
2006/2010	15	2005 (2009), Euro 4	50 (10)

SO_2 to SO_4, and this leads to a significant increase in the mass of the particulate matter and reduction in the efficiency of the aftertreatment system.

2.2.2.8 Aromatics

Increasing the aromatic content will increase the flame temperature during combustion [2.50] and increase NOx emissions [2.16, 2.17]. However, the major effect of aromatics is via polynuclear (multiringed) aromatic hydrocarbons (PAHs). Reducing PAHs reduces HC, CO, and particulate matter (PM) in heavy-duty engines [2.17]. However, modern engines equipped with efficient diesel particulate filter systems are probably not affected much by PAH content [2.17]. In Europe, EN 590 puts a maximum limit of 11% molar concentration on PAHs in diesel fuel.

2.2.2.9 Biodiesel

Biodiesels are fatty acid methyl esters (FAMEs) made by reacting vegetable or animal fat with methanol in the presence of a base and are increasingly being used in diesel fuel to reduce the dependence on petroleum. Rapeseed methyl ester (RME) is the most common biodiesel in Europe. FAMEs are in the diesel boiling range, have CNs in the 50s, and enhance lubricity. However, their energy density is lower than that of conventional diesel because they contain oxygen. Problems could also occur associated with stability, high viscosity at low temperatures, deposit formation in injectors, and incompatibility with fuel system components like rubber seals; they are also hygroscopic, and the increased retention of water could lead to corrosion and microbial growth [2.16]. Their presence in diesel fuel reduces PM because of the presence of oxygen but usually increases NOx [2.16, 2.17]. EN 590 allows blending of up to 7% FAME by volume in European diesel.

2.3 Fuel Additives

Additives are chemicals added to the fuel in small quantities, at concentrations of less than 5000 ppm or 0.5%, to either correct a deficiency or improve performance [2.51]. They are usually manufactured by specialist additive or chemical companies and added at fuel distribution depots by individual oil companies as a package of more than one chemical along with a solvent or diluent like xylene or gasoline. Typical injection rates are 1000–1500 ppm, and up to half the package could be solvent. In general, the dose response is nonlinear; that is, the rate of change of effectiveness of the additive decreases as the additive treat rate is increased, reducing the cost effectiveness of additive use. Responsible fuel and additive companies take great care and invest resources in developing and testing additives before they are commercialized. Extensive engine and road tests are done to collect evidence of the benefits of the additive to support product claims. Equally important is ensuring that no undesirable side effects occur, including inlet valve sticking, catalyst durability, toxicological problems, component incompatibility, handling and distribution problems, lubricant degradation, paint staining, and spark plug fouling [2.52, 2.53, 2.54]. Some types of storage- and distribution-related additives are used in both gasoline and diesel. We now briefly consider some of the more commonly used additives.

2.3.1 Gasoline Performance Additives

2.3.1.1 Deposit Control Additives

Deposit control additives are the most commonly used additives, and in some markets like the United States, gasoline is mandated to contain a minimum amount of such additives. These additives control deposits in fuel injectors and on inlet valves in port fuel-injected engines; controlling these deposits is important (discussed in greater detail in Chapter 3). These surface-active materials consist of a polar head connected via a suitable linking group to a polymer that is soluble in the fuel. The polar head generally contains nitrogen and oxygen hetero atoms that attach themselves to metal surfaces, causing a thin protective film to be formed to prevent deposits sticking to the metal surface.

2.3.1.2 Combustion Modifiers

Antiknock additives, which increase the resistance of the fuel to autoignition, are the most common combustion modifiers used. The most effective of these are organometallic compounds; however, their use, particularly, that of lead compounds, has decreased very significantly in recent times because of concerns over health and their adverse effects on aftertreatment systems. Many of the known nonmetallic or ashless antiknocks are amines or phenols, of which n-methyl aniline (NMA) is among the most effective. However, none of these are cost-effective compared to further processing of the crude to improve gasoline antiknock quality. Many of them also could be toxic or carcinogenic. Therefore, ashless antiknocks are not commonly used in commercial gasoline. Another combustion modifier that was commercially used for a brief period was the spark-aider, which improved ignition and early flame development by improving the efficiency of the spark [2.55]. However, spark-aiders are metallic additives and are not used anymore.

2.3.1.3 Friction Modifiers

Friction modifiers are commonly used in lubricants, but they can also be used as fuel additives to reduce the friction at the piston ring-cylinder wall interface, a location where friction is high and the oil quantity is purposely kept low to improve efficiency [2.56]. Typically, they are polar compounds with long alkyl chains with affinity for metal surfaces.

2.3.2 Diesel Performance Additives

2.3.2.1 Cetane Improvers

Cetane improvers readily decompose to form free radicals that promote autoignition and increase the CN of the fuel. Alkyl nitrates have been commonly used for this purpose because of their low cost. The most commonly used cetane improver is 2-ethylhexyl nitrate (EHN). Peroxides can also act as cetane improvers, but they are more expensive compared to nitrates and are also generally worse for fuel stability. Di-tertiary-butyl-peroxide (DTBP), which was among the first cetane improvers to be commercially recognized, is also used to some extent. The effectiveness of these additives depends on the properties and composition of the base fuel, and empirical correlations to predict the dose response of such additives have been developed [2.57]. At 150 ppm, a typical concentration for EHN use, the CN of the base fuel could be increased by between 2 and 3.5 units.

2.3.2.2 Detergent Additives

Deposits can form inside the injector nozzle or on the injector tip near the exit from the nozzle holes (see Chapter 3 for further discussion). These deposits will reduce the fuel flow rate by reducing the flow area (since the injection pressure is fixed by the engine map at a given operating condition) and thus reduce power. Deposits can also affect the fuel spray pattern, affecting the mixing of fuel and air, and might cause an increase in emissions. Detergent additives are commonly used in diesel fuels to control these deposits [2.58].

2.3.2.3 Cold Flow Performance Additives

The cold weather performance of diesel fuels can be improved through the use of additives [2.3]. Different chemistries that have been shown to be effective are ethylene-vinyl acetate (EVA) polymers, chlorinated hydrocarbons, and polyolefins [2.51]; EVA polymers are the most common type [2.3, 2.59]. Typically, the treatment range is between 100 and 400 ppm, and very difficult fuels require 500 to 1000 ppm. Pour point depressants (PPDs) are usually low-molecular-weight polymers [2.60] and work by inhibiting the growth of wax crystals in one crystallographic plane, thereby producing crystals of higher volume-to-surface-area ratio [2.61]. This also reduces the ability of the crystals to interlock and lowers the PP. They are more commonly used in refinery applications [2.61]; they do not usually improve vehicle operability significantly [2.3].

PPDs are believed to work primarily on wax crystals already formed in the fuel. Cold flow improvers designed to improve vehicle operability (e.g., those that depress CFPP) work through a nucleation mechanism [2.61, 2.62]. Like PPDs, they are low-molecular-weight polymers, but with the n-paraffin-like segments more carefully designed to interact more efficiently with those first n-paraffins that crystallize [2.60]. The additive forms many nuclei before the separation of the wax crystals. When the temperature of the fuel is lowered below the CP, wax crystals can start to form at many more sites compared to the fuel without the additive, creating many small crystals [2.61, 2.62]. Further growth of these crystals is also affected by these additives [2.3, 2.62]. The temperature at which the wax crystals will grow to a size large enough to block the filter in a CFPP rig (i.e., the CFPP temperature) will be lower than for the fuel without the additive. Wax antisettling additives (WASAs) reduce the wax crystal size even further [2.63] and slow down their rate of settling in storage tanks. Some olefin-ester type copolymers have also been shown to reduce CPs [2.64], whereas the conventional cold flow improver additives have little effect on CP [2.3, 2.47]. These additives depress CPs by no more than 3–4 °C even with high treat rates and reduce CFPP by a similar amount, while more conventional flow improvers are far more effective in depressing CFPP [2.3].

2.3.2.4 Lubricity Additives

Low-sulfur diesel fuels usually have to contain lubricity additives to ensure that the high-pressure injection pumps in modern diesel engines are not damaged by excessive wear [2.65]. Lubrication is provided by polar components in the fuel that prevent wear on contacting metal surfaces. These components might be esters, neutral salts, or mono acids. Biodiesels, which are usually fatty acid esters, can significantly increase lubricity. In most diesels, the addition of 2% volume of biodiesel would be sufficient to ensure that the lubricity level is within specifications.

2.3.3 Storage- and Distribution-Related Additives

Storage- and distribution-related additives can be used in both gasoline and diesel fuel. Antioxidants are used to improve the stability, usually of gasoline, in storage by preventing gum formation. Gum is produced by oxidation and polymerization reactions involving mainly the olefinic compounds present in a gasoline. The initial reaction product is a hydroperoxide followed by a series of other reactions giving rise to a complex range of products. Antioxidants function by combining with peroxide free radicals and by decomposing hydroperoxides into stable substances. The most common gasoline antioxidants in use now are aromatic diamines and alkyl phenols. They are used in small concentrations (< 30 ppm). Dehazers are commonly used in gasolines to prevent the formation of emulsions of fuel and water, which might cause haziness of the fuel. Water might enter the fuel during storage or manufacture, and given enough time, will settle out. However, in the presence of detergent additives, the coalescence of water droplets might be inhibited, and it would take longer to clear the fuel of water. Dehazers make water droplets coalesce more easily and are often in detergent packages used for deposit control. Markers are dyes used to identify fuel and prevent theft.

2.4 References

2.1 "Gasoline Explained." U.S. Energy Information Administration. http://www.eia.gov/energyexplained/index.cfm?page=gasoline_environment, Accessed January 9, 2013.

2.2 "History of Biodiesel Fuel." Pacific Biodiesel. http://www.biodiesel.com/index.php/biodiesel/history_of_biodiesel_fuel. Accessed January 9, 2013.

2.3 Owen, K., and Coley, T. 1995. *Automotive Fuels Reference Book*. SAE International, Warrendale, PA.

2.4 "Tetraethyllead." Wikipedia. http://en.wikipedia.org/wiki/Tetraethyllead. Accessed January 9, 2013.

2.5 "2012 The Outlook for Energy: A View to 2040." http://www.exxonmobil.co.uk/corporate/files/news_pub_eo2012.pdf. Accessed June 9, 2013.

2.6 Speight, J.G. 2007. "Petroleum Refinery Processes." In *Wiley Critical Content: Petroleum Technology, Vol. 1*. John Wiley & Sons, Hoboken, NJ.

2.7 Heywood, J.B. 1988. *Internal Combustion Engine Fundamentals*. McGraw-Hill, New York.

2.8 "Gas to Liquids." http://www.chemlink.com.au/gtl.htm. Accessed January 9, 2013

2.9 *Annual Book of ASTM Standards, Vol. 5.01–5.05*. 2013. ASTM International, West Conshohocken, PA.

2.10 *IP Standard Test Methods for Analysis and Testing of Petroleum and Related Products and British Standard 2000 Parts. 2013*. EI, London.

2.11 CEN: European Committee for Standardization. http://www.cen.eu/cen/pages/default.aspx. Accessed January 9, 2013.

2.12 "Fuels and Fuel Additives." U.S. EPA. http://www.epa.gov/otaq/fuels/index.htm. Accessed January 9, 2013.

2.13 Burns, R.V., Benson, J.D., Hochhauser, A.M., Koehi, W.J., Kreucher, W.M., and Reuter, R.M. 1991. "Description of Auto/Oil Air Quality Research Programme." SAE Paper No. 912320. SAE International, Warrendale, PA.

2.14 Mackinven, R., and Hublin, M. 1996. "European Programme on Emissions, Fuels and Engine Technologies—Objectives and Design." SAE Paper No. 961065. SAE International, Warrendale, PA.

2.15 Tsuda, H., Ito, T., and Nakamura, K. 1999. "Japan Clean Air Program (JCAP)—Program Objectives and Design." SAE Paper No. 1999-01-1481. SAE International, Warrendale, PA.

2.16 "Worldwide Fuel Charter." OICA. http://oica.net/worldwide-fuels-charter/. Accessed January 9, 2013.

2.17 Hochhauser, A.M. 2008. *Review of Prior Studies on Fuel Effects on Vehicle Emissions*. CRC Report E-84, Coordinating Research Council, August 2008. Also published as SAE Paper 2009-01-1181. 2009. SAE International, Warrendale, PA.

2.18 Reuter, R.M., Benson, J.D., Burns, R.V., Gorse, R.A., Hochhauser, A.M., Koehl, W.J., Painter, L.J., Rippon, B.H., and Rutherford, J.A. 1992. "Effects of Oxygenated Fuels and RVP on Automotive Emissions—Auto/Oil Air Quality Improvement Program." SAE Paper No. 920326. SAE International, Warrendale, PA.

2.19 Koseki, K., Matsumoto, T., Morita, K., Hirose, M., and Sembokuya, S. 2000. "RVP Dependence of Evaporative Emissions for Japanese Current and Older Vehicles and U.S. Vehicles Using Typical Japanese Gasoline." SAE Paper 2000-01-1170. SAE International, Warrendale, PA.

2.20 Jorgensen, S.W., Musser, G.S., Uihlein, J.P., and Evans, B. 1996. "A New CRC Cold-Start and Warm-Up Driveability Test and Associated Demerit Weighting Procedure for MPFI Vehicles." SAE Paper No. 962024. SAE International, Warrendale, PA.

2.21 Stephenson, T., and Paesler, H. 1996. "The Development of a Cold Weather Driveability Test Cycle for Fuel Injected Vehicles." SAE Paper No. 96122. SAE International, Warrendale, PA.

2.22 Jewitt, C.H., Gibbs, L.M., and Evans, B. 2005. "Gasoline Driveability Index, Ethanol Content and Cold-Start/Warm-Up Vehicle Performance." SAE Paper No. 2005-01-3864. SAE International, Warrendale, PA.

2.23 Geng, P.Y., Lund, V.A., Studzinski, W.M., Waypa, T.G., and Belton, D.N. 2007. "Gasoline Distillation Effect on Vehicle Cold Start Driveability." SAE Paper No. 2007-01-4073. SAE International, Warrendale, PA.

2.24 Jorgensen, S.W., and Benson, J.D. 1994. "Simultaneous Measurements of Driveability and Emissions at Cool Ambient Temperatures." SAE Paper No. 941870. SAE International, Warrendale, PA.

2.25 Kuk, K., Kwan, Y., Brown, M., and Schmidt, M. 2000. "The Effects of Driveability on Emissions in European Gasoline Vehicles." SAE Paper No. 2000-01-1884. SAE International, Warrendale, PA.

2.26 Gorse, R.A., Benson, J.D., Burns, V.R., Hochhauser, A.M., Koehl, W.J., Painter, L.J., Reuter, R.M., and Rippon, B.H. 1991. "Toxic Air Pollutant Vehicle Exhaust Emissions with Reformulated Gasolines." SAE Paper No. 912324. SAE International, Warrendale, PA.

2.27 Goodfellow, C.L., Gorse, R.A., Hawkins, M.J., and McArragher, J.S. 1996. "European Programme on Emissions, Fuels and Engine Technologies

(EPEFE)—Gasoline Aromatics/E100 Study." SAE Paper No. 961072. SAE International, Warrendale, PA.

2.28 EPA. 2008. *Summary and Analysis of the 2008 Gasoline Benzene Pre-Compliance Reports*. EPA 420-R-08-022, December 2008. U.S. EPA.

2.29 Schleyer, C.H., Koehl, W.J., Leppard, W.R., Dunker, A.M., Yarwood, G., Cohen, J.P., and Pollack, A.K. 1994. "Effect of Gasoline Olefin Composition on Predicted Ozone in 2005/2010—Auto/Oil Air Quality Improvement Research Program." SAE Paper No. 940579. SAE International, Warrendale, PA.

2.30 Benson, J.D., Burns, V., Koehl, W.J., Gorse, R.A., Painter, L.J., Hochhauser, A.M., and Reuter, R.M. 1991. "Effects of Gasoline Sulfur Level on Mass Exhaust Emissions—Auto/Oil Air Quality Improvement Research Program." SAE Paper No. 912323. SAE International, Warrendale, PA.

2.31 Petit, R.A., Jeffery, J.G., Palmer, F.H., and Steinbrink, R. 1996. "European Programme on Emissions, Fuels and Engine Technologies (EPEFE)—Emissions from Gasoline Sulphur Study." SAE Paper No. 961071. SAE International, Warrendale, PA.

2.32 Schleyer, C.H., Gorse, R.A., Gunst, R.F., Barnes, G.J., Eckstrom, J., Eng, K.D., Natarajan, M., and Schlenker, A.M. 1998. "Effect of Fuel Sulfur on Emissions in California Low Emission Vehicles." SAE Paper No. 982726. SAE International, Warrendale, PA.

2.33 Koseki, K., Uchiyama, T., Kawamura, M., and Sembokuya, S. 2000. "A Study on the Effects of Sulfur in Gasoline on Exhaust Emissions." SAE Paper No. 200-01-1878. SAE International, Warrendale, PA.

2.34 Hochhauser, A.M., Schleyer, C.H., Yeh, L.I., and Rickeard, D.J. 2006. "Impact of Fuel Sulfur on Gasoline and Diesel Vehicle Emissions." SAE Paper No. 2006-01-3370. SAE International, Warrendale, PA.

2.35 International Petroleum Industry Environmental Conservation Association (IPIECA). 2006. *Fuel Sulphur: Strategies and Options for Enabling Clean Fuels and Vehicles*. http://www.ipieca.org/sites/default/files/publications/Sulphur.pdf. Accessed January 9, 2013.

2.36 "Fuels and Additives: E15 (A Blend of Gasoline And Ethanol)." U.S. EPA. http://www.epa.gov/otaq/regs/fuels/additive/e15/. Accessed January 9, 2013.

2.37 "European Biofuels Technology Platform, Biobutanol." http://www.biofuelstp.eu/butanol.html. Accessed January 9, 2013.

2.38 Durbin, T.D., Miller, J.W., Younglove, T., Huai, T., and Cocker, K. 2006. "Effects of Ethanol and Volatility Parameters on Exhaust Emissions CRC Project No. E-67." Coordinating Research Council.

2.39 California Air Resources Board. 1999. "Air Quality Impacts of the Use of Ethanol in California Reformulated Gasoline." UCRL-AR-135949, Vol. 3, December 1999.

2.40 Benson, J.D., and Dana, G. 2002. "The Impact of MMT Gasoline Additive on Exhaust Emissions and Fuel Economy of Low Emission Vehicles (LEV)." SAE Paper No. 2002-01-2894. SAE International, Warrendale, PA.

2.41 Cunningham, L.J., Hollrah, D.P., Newkirk, M.S., and Roos, J.W. 2003. "AAM/AIAM Fleet Test Program: Analysis and Comments." SAE Paper No. 2003-01-3287. SAE International, Warrendale, PA.

2.42 "Fuels and Fuel Additives: Comments on the Gasoline Additive MMT." U.S. EPA. http://www.epa.gov/otaq/regs/fuels/additive/mmt_cmts.htm. Accessed on January 9, 2013.

2.43 Kalghatgi, G.T. 2005. "Auto-Ignition Quality of Practical Fuels and Implications for Fuel Requirements Of Future SI and HCCI Engines." SAE Paper No. 2005-01-0239. SAE International, Warrendale, PA.

2.44 Ryan, T.W., and Matheaus, A. 2002. "Fuel Requirements for HCCI Engine Operation." THIESEL 2002, Thermo and Fluid Dynamic Processes in Diesel Engines.

2.45 "European Programme on Emissions, Fuels and Engine Technologies (EPEFE), Final Report." 1995. ACEA/EUROPIA.

2.46 Barbour, R.H., Rickerad, D.J., and Elliott, N.G. 2000. "Understanding Diesel Lubricity," SAE Paper No. 2000-01-1918. SAE International, Warrendale, PA.

2.47 Reddy, S.R., and McMillan, M.L. 1981. "Understanding the Effectiveness of Diesel Fuel Flow Improvers." SAE Paper No. 811181. SAE International, Warrendale, PA.

2.48 McMillan, M.L., and Barry, E.G. 1983. "Fuel and Vehicle Effects on Low-Temperature Operation of Diesel Vehicles—The 1981 CRC Field Test." SAE Paper No. 830594. SAE International, Warrendale, PA.

2.49 Chandler, J.E., Horneck, F.G., and Brown, G.I. 1992. "The Effect of Cold Flow Additives on Low Temperature Operability of Diesel Fuels." SAE Paper No. 922186. SAE International, Warrendale, PA.

2.50 Takahashi, S., Wakimoto, K., Iida, N., and Nikolic, D. 2001. "Effects of Aromatics Content and 90% Distillation Temperature of Diesel Fuels on Flame Temperature and Soot Formation." SAE Paper No. 2001-01-1940. SAE International, Warrendale, PA.

2.51 Tupa, R.C., and Dorer, C.J. 1986. "Gasoline and Diesel Fuel Additives for Performance/Distribution Quality—II." SAE Paper No. 861179. SAE International, Warrendale, PA.

2.52 Koehler, D.E., and Doglio, J.A. 1987. "Benefits of Multifunctional Diesel Fuel Additives Demonstrated in a Fleet Test." SAE Paper No. 872146. SAE International, Warrendale, PA.

2.53 Spink, C.D., McDonald, C.R., Morris, G.E.L., and Stephenson, T. 2004. "A Critical Road Test Evaluation of a High-Performance Gasoline Additive Package in a Fleet of Modern European and Asian Vehicles." SAE Paper No. 2004-01-2027, 2004. SAE International, Warrendale, PA.

2.54 Spink, C.D., Cantlay, A.J., Ziman, P.R., Leubbers, M., and Lance, D. 2006. "A New Protocol for the Road Test Evaluation of Gasoline Additive Packages." SAE Paper No. 2006-01-3408. SAE International, Warrendale, PA.

2.55 Kalghatgi, G.T. 1988. "Effect of Spark-Aider Fuel Additives on the Misfire Characteristics of a Spark Ignition Engine." *Combustion Science and Technology* 62:1–19.

2.56 Hayden, T.E., Ropes, C.A., and Rawdon, M.G. 2001. "The Performance of a Gasoline Friction Modifier Fuel Additive." SAE Paper No. 2001-01-1961. SAE International, Warrendale, PA.

2.57 Thompson, A.A., Lambert, S.W., and Mulqueen, S.C. 1997. "Prediction and Precision of Cetane Number Improver Response Equations." SAE Paper No. 972901. SAE International, Warrendale, PA.

2.58 Barbour, R., Arters, G., Dietz, J., Macduff, M., Panesar, A., and Qugley, R. 2007. "Diesel Detergent Additive Responses in Modern, High-Speed, Direct-Injection, Light-Duty Engines." SAE Paper No. 2007-01-2001. SAE International, Warrendale, PA.

2.59 Gairing, M., Marriott, J.M., Reders, K.H., Reglitzky, A.A., and Wolveridge, P.E., "The Effect of Modern Additive Technology on Diesel Fuel Performance," SAE Paper No. 950252. SAE International, Warrendale, PA.

2.60 Chandler, J.E., Horneck, F.G., and Brown, G.I. 1992. "The Effect of Cold Flow Additives on Low Temperature Operability of Diesel Fuels." SAE Paper No. 922186. SAE International, Warrendale, PA.

2.61 Botros, M.G. 1997. "Enhancing the Cold Flow Behavior of Diesel Fuels." SAE Paper No. 972899. SAE International, Warrendale, PA.

2.62 Zielinski, J., and Rossi, F. 1984. "Wax and Flow in Diesel Fuels." SAE Paper 841352. SAE International, Warrendale, PA.

2.63 Brown, G.I., Tack, R.D., and Chandler, J.E. 1988. "An Additive Solution to the Problem of Wax Settling in Diesel Fuels," SAE Paper No. 881652. SAE International, Warrendale, PA.

2.64 Damin, B., Faure, A., Denis, J., Sillion, B., Claudy, P., and Letoffe, J.M. 1986. "New Additives for Diesel Fuels: Cloud Point Depressants." SAE Paper No. 861527. SAE International, Warrendale, PA.

2.65 Batt, R.J., McMillan, J.A., and Bradbury, I.P. 1996. "Lubricity Additives—Performance and No-Harm Effects in Low Sulfur Fuels." SAE Paper No. 961943. SAE International, Warrendale, PA.

Chapter 3
Deposits in Internal Combustion Engines

Deposits form on all the internal surfaces of an engine with use and can affect engine performance and emissions significantly. Mostly, these effects are for the worse and not quantitatively predictable, so engine designers cannot take them into account to optimize their designs. In some cases, deposits can cause engine damage. Engine deposits are classified into three broad categories: fuel system deposits, intake system deposits, and combustion chamber deposits. In both spark ignition (SI) and compression ignition (CI) engines, the deposits of most concern in the fuel system occur in the fuel-metering devices, such as the fuel injector or, in older SI engines, the carburetor. Modern SI engines do not use carburetors, and carburetor deposits will not be considered further in this chapter (for more information, see [3.1]). Combustion chamber deposits can cause problems in SI engines, and intake system deposits can cause problems in SI engines with port fuel injection (PFI).

Figure 3.1a shows a sketch of the cross section of a diesel injector. The injector needle is actuated by a solenoid valve and moves to allow the fuel to flow under pressure through the nozzle holes into the combustion chamber. Deposits can build up on the internal surfaces of the nozzle hole to reduce the available area and hence the fuel flow rate at a given fuel pressure. These actual deposit weights may be very small (10–50 μg), but since the orifice diameters are very small, the deposits could have a significant effect on the fuel flow rate. A 10-μm thick deposit layer in a 150-μm diameter circular hole (diesel injector holes are much smaller) would reduce the flow area by 25%; flakes of deposits up to 15-μm thick have been found in port fuel injectors in the field [3.2]. Figure 3.1b shows deposits near the nozzle hole of a diesel injector from a passenger car, and Fig. 3.1c shows a magnified view of such deposits. In recent years, deposits on the internal surfaces of a diesel injector (e.g., on the injector needle (Fig. 3.1d) or at the interface of the needle and the nozzle body) have been causing concern [3.3–3.6], particularly in heavy-duty engines in North America. Such deposits can slow the response of the injector by causing the needle to stick and cause rough running, higher emissions, and a reduction in efficiency.

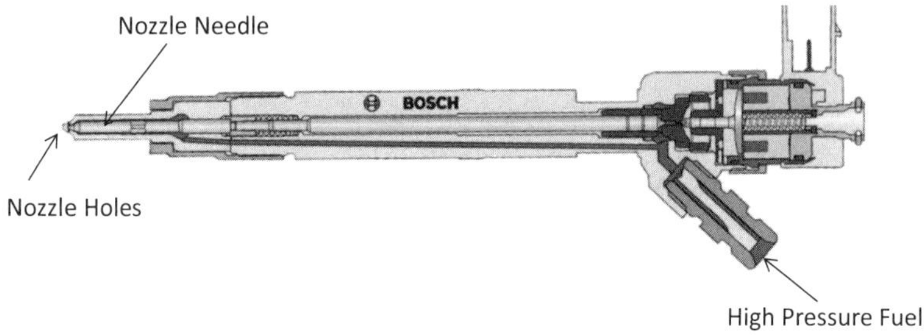

Fig. 3.1a. Diesel injector cross section. (Copyright by Robert Bosch GmBH. Reproduced with permission.)

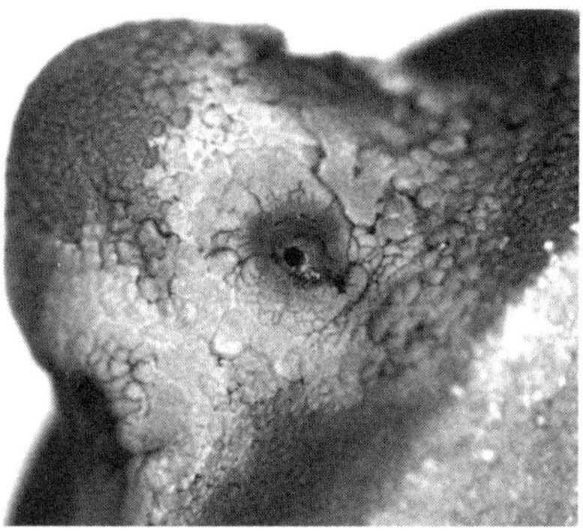

Fig. 3.1b. Deposits at a diesel injector tip. (Copyright by Afton Chemical Ltd. Reproduced with permission.)

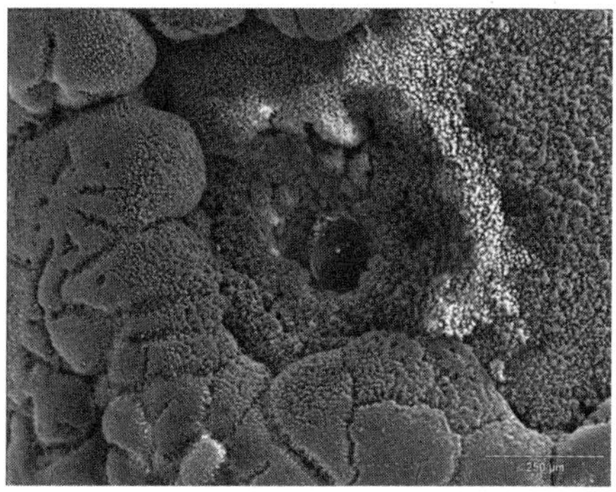

Fig. 3.1c. Micrograph of diesel injector deposits near the nozzle hole. (Copyright by Afton Chemical Ltd. Reproduced with permission.)

Fig. 3.1d. Diesel injector internal deposits: Nozzle needle with deposits consisting of polymeric organic material with additive related constituents. The scratches come from a needle used to remove material for analysis. (Copyright by Robert Bosch GmBH. Reproduced with permission.)

Figure 3.2 shows a slit injector from a direct injection spark ignition (DISI) engine after the vehicle had been run for 15,000 miles on the road. Deposits also build up on the outside surfaces around the orifices and can upset the fuel flow pattern. Figure 3.3 compares the nozzle hole of a diesel injector before and after the deposits have been built up. Figure 3.4 illustrates how deposits could upset the fuel spray pattern in a diesel engine.

Figure 3.5 shows the position of deposits in different parts of an SI engine with PFI.

The physical and especially the chemical mechanisms that are involved in deposit formation are very complex, differ in different parts of the engine because of the differences in the chemical and thermal environments, and are not fully understood. The mechanisms through which deposits affect engine performance and emissions are also complex, and the effects themselves

Fig. 3.2. Slit injector from a DISI engine after the vehicle had run on the road for 15,000 miles on an unleaded gasoline without additives (Copyright by Shell Global Solutions. Reproduced with permission.)

NEW INJECTOR FOULED INJECTOR

Fig. 3.3. Electron micrographs of the combustion chamber end of the nozzle hole of a diesel injector—with and without deposits: new injector, fouled injector. (Copyright by Shell Global Solutions. Reproduced with permission.)

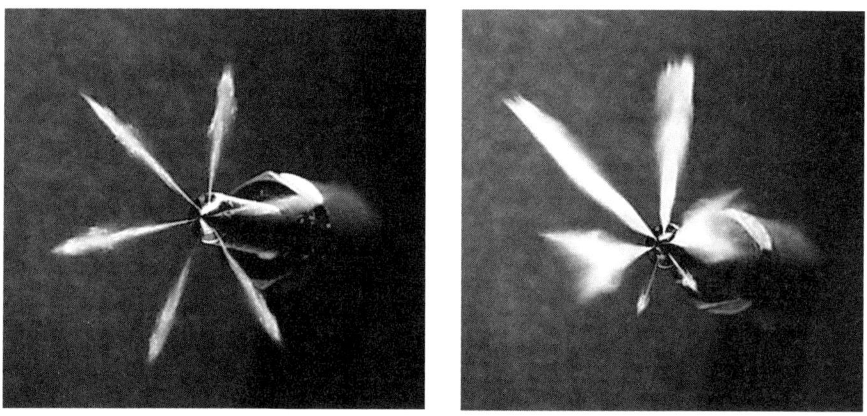

Fig. 3.4. Effects of injector deposits on the spray pattern of a diesel injector spray. The clean injector is on the left. The injector with deposits on the right shows very uneven spray pattern. (Copyright by Shell Global Solutions. Reproduced with permission.)

can only be quantified with extensive testing. These effects are different in different engines and at different operating conditions in the same engine. Large-scale fleet tests and rigorous statistical analysis are always necessary to estimate deposit-related effects reliably across a wide range of vehicles. Engine and vehicle tests have been used extensively to study deposits in engines. However, these tests are characterized by poor repeatability and

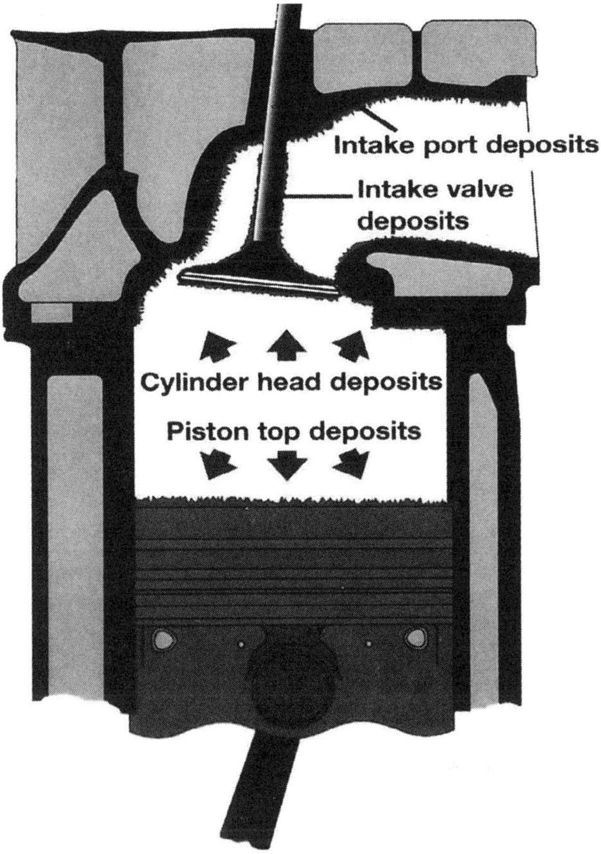

Fig. 3.5. Sketch showing deposits in a PFI SI engine.

reproducibility because of the complexity of the mechanisms involved and the impossibility of controlling all the test conditions accurately over the relatively long time periods such tests have to run. Moreover, the parameters used to measure the severity of the problem, such as valve ratings, might be qualitative and determined subjectively. The tests are also usually expensive and slow. Cheaper and quicker laboratory and rig tests have been developed but usually do not adequately reflect the experience in the field. In general, deposit levels reflect the equilibrium between deposit formation and deposit removal mechanisms that take place simultaneously.

This chapter will discuss the nature, formation, and effects of deposits in both SI and diesel engines. It will also consider the measures employed to control or reduce them.

3.1 The Nature and Formation of Deposits

Thermal degradation and oxidation of the fuel and lubricant play an important role in the formation of deposits in the intake system and the injectors. The literature review by Hilden [3.7] highlights the complexity of the processes involved in the oxidation of gasoline. Many of the components that are important in deposit formation, such as polar compounds, only exist at very low concentrations. It is usually not possible to identify all such potentially reactive materials. The oxidation products of other components, also present at low concentrations, might also be deposit precursors [3.8]. The presence of metals, and compounds containing oxygen, nitrogen, and sulfur, is known to affect the deposit-forming tendency of gasolines [3.9, 3.10]. However, these effects can vary widely—from vastly enhancing deposit formation to inhibiting it—depending on the detailed chemistry involved and the composition of the gasoline; such variation has been demonstrated for thioanisole, a sulfur compound [3.9]. The presence of trace levels of zinc and copper in the fuel can significantly enhance diesel injector fouling [3.11]. Zinc, in particular, appears to enhance the formation of deposit precursors as well as contribute directly to deposit formation, given that it is detected in actual deposits in diesel injectors [3.12, 3.13]. Some major fuel components have been shown to be bad for oxidation stability and deposit formation. Kolobielski and McCaleb ([3.14], quoted in [3.7]) suggest the following general ranking for decreasing gum formation propensity: diolefins, aromatic olefins, olefins, aromatics, isoparaffins, naphthenes, and paraffins. However, the relative order may change depending on the chemical structure of the actual compound and the conditions at which the deposits form. The interaction between metal surfaces and deposit precursors is also likely to be important and is largely unexplored. For instance, lower amounts of deposits were found to form on steel surfaces covered by thick oxide films than on clean metal [3.9].

In general, the deposit level reflects the balance between deposit formation and removal mechanisms that are active simultaneously. At any given condition, this level will reach equilibrium (i.e., deposits do not grow indefinitely). Clearly, conditions under which deposits form, such as surface temperatures and the processes that occur round these surfaces, are important. For instance, injector tips can experience temperatures of over 100 °C in PFI engines [3.2, 3.10, 3.15], up to 180 °C in DISI engines [3.16], and more than 300 °C in diesel engines [3.13]. The temperatures on the intake port and valve surfaces can reach up to 350 °C [3.17], while the average surface temperatures in the combustion chamber vary widely across the combustion chamber, from around 120 °C [3.18, 3.19] in the cooler areas to about 320 °C at the exhaust valve seats to over 800 °C at the exhaust valves [3.20]. The blowby gases from the gaps between the piston, piston rings, and cylinder walls escape into the crankcase and are returned to the intake system in modern engines by the PCV (positive crankcase ventilation) system in order to reduce hydrocarbon emissions. It is also common to recirculate a part of the exhaust gas to the inlet system to reduce NOx emissions. These EGR (exhaust gas recirculation) gases are hot; contain particulates, products of combustion, and deposit precursors; and can increase deposit formation in port fuel injectors and intake valves [3.21, 3.22]. The mechanism of deposit formation in the combustion chamber differs significantly from that in the fuel and intake systems.

3.1.1 Fuel Injector Deposits

3.1.1.1 Port Fuel Injector Deposits

When multiport fuel injector (MPFI) SI engines were first introduced in the United States in the 1980s, there was a substantial problem associated with injector deposits that led to a large effort aimed at understanding the nature and formation of PFI deposits [3.2, 3.7–3.10, 3.15, 3.23–3.28]. PFI deposits do not form if the engine is operated continuously—either at high speed or at idle [3.15]. A period of hot soak when the engine is fully warmed up but not running, during which the tip temperature has been measured to reach 100 °C, is essential for PFI deposit formation [3.10, 3.24]. Taniguchi et al. [3.10] found that the operating cycle that produced the most PFI deposits consisted of a short (15-minute) cruise followed by a long (45-minute) soak. Deposit levels decreased considerably when the ratio of cruise times to soak times was increased either by increasing the cruise period or decreasing the soak period. With increased cruise period, the deposit precursors would have a greater chance of being removed by the fuel. This observation is in line with the finding in [3.15] that a high-duty cycle, with a higher average fuel flow rate, produced less deposit than a low-duty cycle. If the soak period is short, deposit precursors are not given enough time to polymerize and can be washed off more easily [3.10, 3.15]. Any design or operating feature that influences the hot-soak temperature affects PFI deposit formation [3.23, 3.25]. Thus, there are significant differences among car makes [3.27].

Thus the mechanism of PFI deposit formation appears to be as follows [3.2, 3.23]: When the engine stops, there is always a small amount of fuel left at the injector tip. During the hot-soak period, the lighter hydrocarbons evaporate, leaving a thin film rich in heavy hydrocarbons on the orifice surfaces. These hydrocarbons oxidize to form gums and resins, and this sticky varnish can bake into a hard deposit given sufficient time. It can also bind particulates such as airborne dirt and solids from combustion products brought into the intake by EGR and PCV systems. However, such combustion products are not essential. PFI deposits could be produced by soaking the injectors in the fuel at 90 °C outside an engine for a sufficient length of time [3.2, 3.15, 3.28]. The deposits are made up of both organic and inorganic material. They consist of a thin, mostly carbonaceous (60–85% carbon), varnish-type under-layer and a modular, brittle layer with larger concentrations of inorganic materials [3.2, 3.10]. The organic fraction consists mainly of partially oxidized hydrocarbons, probably originating from the fuel, whereas the inorganics are largely made up of sulfates and oxides of metals [3.2, 3.10, 3.15] presumably derived from the lubricant and the injector body. The overall composition can vary greatly; deposits that were particularly difficult to remove showed exceptionally large fractions (~75%) of inorganic material [3.10]. High olefin levels in the base fuel increase PFI deposit formation [3.2, 3.10, 3.15, 3.27, 3.28], whereas aromatics appeared to be unimportant [3.2]. Conjugated diolefins such as 1,3-pentadiene and 1,3-butadiene were found to induce deposit formation when added in low concentrations to "clean" fuels [3.2]. Polar materials and nitrogen and sulfur compounds [3.10] have also been implicated in PFI deposit formation. Laboratory tests showed that deposit formation was reduced when diolefins [3.7] and polar compounds [3.8] were removed from an unstable gasoline. Conversely, when a mixture of diolefins was added to a stable gasoline containing olefins, PFI deposit formation was increased, but the same mixture had little effect when added to a stable gasoline with little olefin content [3.7]. Thus the effect of diolefins or olefins on PFI deposit formation depends on what the base fuel contains [3.7, 3.10]. In general, conventional fuel stability criteria do not correlate with PFI deposit formation [3.15].

3.1.1.2 Direct Injection Spark Ignition Injector Deposits

Unlike PFI deposits, DISI injector deposits form even when the engine is run continuously [3.16, 3.29]. The injector tip temperature has a very strong influence on DISI injector formation—injector plugging gets worse as tip temperature increases to about 170 °C. At a slightly higher tip temperature of 184 °C, injector plugging, as measured by percent flow loss, decreased significantly for all the ten fuels tested in [3.16]. At a fixed load, increasing the speed has little effect on the tip temperature, but tip temperature increases as the load is increased for a given speed [3.30]. It has been suggested [3.29] that there should always be liquid fuel at the tip so that deposit precursors could be washed away by fuel injections. Hence DISI injector deposits can be minimized if the tip temperature is below T90, the temperature at which 90% of the fuel is vaporized. Indeed, it has been found that T90 is the most important fuel parameter, with lower T90 leading to higher levels of deposit formation [3.29, 3.30]. Higher sulfur or olefin content also leads to higher levels of DISI injector deposits [3.29]. The reduction in injector plugging on increasing the tip temperature from about 170 °C, as reported in [3.16], might be a result of the higher fuel flow rate at the higher load associated with the higher tip temperature and hence more efficient removal of deposit precursors. The deposit levels again reflect a balance between deposit formation and removal by liquid fuel, as in PFI deposits, except that the tip temperature in DISI injectors can be much higher than in PFI. The tips of DISI injectors as well as those of diesel injectors [3.31–3.39] are exposed to the gases and processes inside the combustion chamber. The mechanisms involved in combustion chamber deposit formation [3.40], as discussed in Section 3.1.3, can also be expected to play a role in deposit formation in such injectors.

3.1.1.3 Diesel Injector Deposits

Diesel injector deposits were more of a problem in indirect injection (IDI) engines operating with pintle-type injector nozzles compared to the direct injection (DI) engines that followed [3.31]. However, even small amounts of deposits can be problematic as injector hole sizes decrease. Moreover, small deviations in spray pattern could affect combustion and emissions significantly in modern engines, in which the mixing of fuel with air needs to be precisely controlled.

Zinc is an extremely important factor in deposit formation in modern diesel engines [3.13, 3.31–3.34]. In fact, the European test protocol used to study diesel injector deposits, the Peugeot DW10 procedure, requires that 1 ppm of zinc, in the form of zinc neodecanoate, be added to the base fuel [3.35]—otherwise deposit formation does not occur at a measurable level or with sufficient repeatability. Metals like zinc and copper can build up in the fuel carried in a vehicle with time, even when the fuel does not contain these metals initially when it is first put into the vehicle [3.36, 3.37]. Ikemoto et al. [3.34] suggest that zinc carboxylate forms close to the nozzle hole outlet by reactions between the zinc in the fuel and lower carboxylic acid in the combustion gas in the initial phase of deposit formation. Subsequently, the main component changes to zinc carbonate close to nozzle hole inlet by reactions with CO_2 in the combustion gas. Metal components and combustion gases are essential elements in the formation of these deposits. Analysis of the deposits shows zinc to be present in different oxidation states [3.13, 3.32, 3.34]. Carbon, oxygen, calcium, and sulfur (assumed to be from the lubricant) are also detected [3.13, 3.32].

The structure of these deposits has been discussed in [3.13, 3.32]. Biodiesel is normally considered to be worse than conventional diesel for deposit formation [3.33]. However, the fouling propensity of biodiesels varies from sample to sample, and it is not clear why this should be so [3.31, 3.38]. FAMEs (fatty acid methyl esters) available in the European market did not show any significant injector deposit formation when blended into a conventional diesel fuel at a level of 10% by volume [3.39]; however, these tests were done without adding zinc to the fuels. Increased nozzle tip temperature leads to increased deposit formation [3.13]. Cavitation inside the nozzle holes produce high local pressures as the micro-bubbles burst and can remove deposit precursors [3.13, 3.33, 3.34]. Nozzles with sharp edges at the inlet, which are more likely to produce cavitation, are less susceptible to injector plugging [3.13]. The physical and especially the chemical processes involved in deposit formation are not well understood, but the residual fuel left in the nozzle at the end of each injection is again considered to be essential to start the process [3.33]. The temperature of this fuel depends on the fuel temperature in the tank, the temperature of the combustion walls, and the protrusion of the nozzle into the combustion chamber.

Internal injector deposits are not exposed to any combustion gases, and their formation appears to be governed by the interaction between the fuel and additives in the fuel, though the detailed mechanisms are again not understood [3.3–3.6]. Two different types of internal deposits have been identified. The first type consists of waxy metal soaps rich in sodium [3.4, 3.5], which are believed to come from sources such as refinery salt driers, pipeline corrosion inhibitors, and storage tank water bottoms. The second type consists of sticky brown deposits postulated to arise from reaction between some deposit-control additives and acidic fuel components [3.5, 3.6]. The internal surface temperatures in the injector are much lower than at the tip, and deposit precursors are unlikely to be formed in situ. One can speculate that cavitation upstream of the injector (e.g., in the fuel pump) could be a source of such precursors, because extremely high local temperatures can be produced as the micro-bubbles produced by cavitation burst.

3.1.2 Intake System Deposits

Intake system deposits—those in the intake port and on the backs of intake valves—are of concern in PFI engines. They also form in DISI engines and diesel engines, but generally do not affect engine performance since they have no direct interaction with the fuel entering the engine. In some first-generation DISI engines, intake valve deposits (IVDs) were so heavy that they affected intake flow patterns. In direct injection stratified charge (DISC) engines mixture formation is critically dependent on intake flow. Hence, such heavy deposits can cause problems in DISC engines, but deposit levels are usually not so heavy in modern DISC engines.

Deposits on the valve tulip can range in weight from fractions of a gram to several grams in some older carburetor engines [3.41, 3.42]. The nature of the deposits can vary from soft and oily to sticky to hard and cokelike [3.42, 3.43] depending on the fuel and additive used and the operating conditions. Figure 3.6 shows a particularly dirty valve and the structure of such deposits under magnification; it is rare to come across such heavy deposits in modern engines. The composition of the deposits depends on the fuel and additives used. For instance, when lead was present in fuels, 30–50% of the deposit weight could be lead [3.42]. Carbon, oxygen, and nitrogen are always present [3.17, 3.42, 3.44]. The oxygen (10–15%) and nitrogen (2–4%) concentrations are very much higher than in the fuel or the lubricant [3.17, 3.42]. The deposits also contain elements such as calcium, zinc, and phosphorous, clearly derived from oil [3.17, 3.45]. The concentrations of these metals are 10–30 times higher than in the original oil. The following four major components have been identified [3.44] in valve deposits in an engine running on unleaded gasoline: monomeric hydrocarbons, carbon black, metal salts and oxides, and a polymer fraction that could act as a binder and promote deposit accumulation.

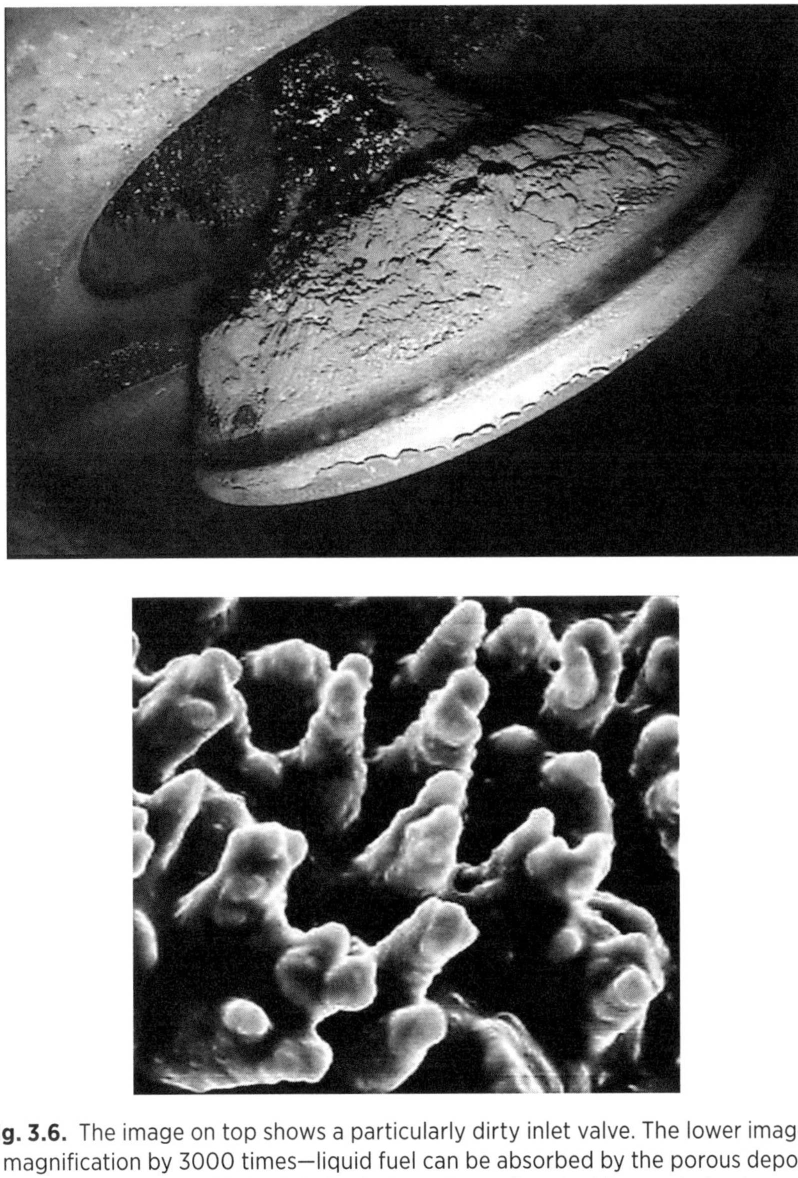

Fig. 3.6. The image on top shows a particularly dirty inlet valve. The lower image is a magnification by 3000 times—liquid fuel can be absorbed by the porous deposit. (Copyright by Shell Global Solutions. Reproduced with permission.)

Liquid hydrocarbons from the fuel and some from the oil are always present on the valve tulips, and some of these can be squeezed between the valve seat and forced back on the tulip as revealed by optical studies [3.42]. These liquid fractions interact with the blowback and EGR gases to participate in deposit formation, but they are also important in the removal of deposits by washing action [3.17, 3.42, 3.46]. Therefore, when the oil flow down the valve stem is increased, valve deposits increase initially [3.27, 3.42, 3.47], but at higher oil flow rates, they decrease [3.42, 3.46, 3.48]. Valve deposits appear to build up quickly—the first 1000 miles or so in vehicles on the road [3.47] or during the first 100 hours or so in engine tests [3.49]—and then fluctuate around more or less stabilized values.

Higher olefin content in the fuel gives rise to higher IVD levels [3.50, 3.51]. Deposits formed by olefin-rich fuels contained larger fractions soluble in n-hexane and acetone [3.50] when compared to those formed with fuels with high aromatic content; deposit weights did not correlate with aromatic content of the fuel. Alcohols [3.52–3.55] and cat-cracked components in the fuel have been found to increase deposit levels. Thermal stability of the fuel and additives in the fuel in the vicinity of the inlet valves and very important if IVDs are to be controlled. Indeed, additives that worked in keeping carburetors clean significantly increased IVDs [3.10, 3.15, 3.42, 3.47]. Similarly, viscosity index (VI) improver additives in the engine oil [3.56], corrosion inhibitors [3.57] used in E85 (85% ethanol and 15% gasoline by volume) and antioxidants in the fuel [3.58] have been shown to cause an increase in IVDs.

The valve and port surface temperatures are very important in deposit formation. The average valve temperature remains constant during an engine cycle, but it changes with engine operating conditions [3.42, 3.59]. When the average valve temperature was reduced by lowering the coolant temperature so that it was below 200 °C 60% of the time instead of 20% of the time, IVD levels halved [3.59]. Higher load and speed conditions lead to increased levels of IVDs [3.41, 3.49] presumably because of higher valve temperatures. However, when the valve temperature was very high (over 360 °C), virtually no deposits were formed in a 60-hour constant speed test [3.59]. Daneshgari and coworkers [3.59, 3.60] explained the temperature effect in terms of the evaporation characteristics of a fuel droplet on a hot surface. At low temperatures, a fuel droplet wets the surface and evaporates slowly. As the temperature increases, its lifetime decreases; it decreases especially rapidly beyond the boiling point. However, if the surface temperature is increased further, above the Leidenfrost temperature, T_L, the droplet will be in the film boiling state and will not wet the surface. A gasoline is made up of many components with a range of T_L. Daneshgari [3.60] suggested that the highest T_L for gasolines is above 250 °C. In general the deposition rate increases as the surface temperature is increased, but if the surface temperature is above T_L, the deposition rate is reduced. Esaki et al. [3.17] suggested that the critical temperature range for IVD formation is between 230 and 350 °C.

Bitting et al. found no effect of different valve surface treatments on IVD formation [3.52]. Similarly, in a rig test, the deposition rate did not depend on the surface material [3.60]. The chemical mechanisms involved in IVD formation are complex and not well understood. Several mechanisms and reaction pathways are believed to occur simultaneously [3.7, 3.44, 3.61, 3.62]. Reactive hydrocarbons may be induced by chemical attack or thermal decomposition to form new bonds with other molecules. Further polymerization and loss of volatile components also play a role [3.62]. Polar hydrocarbons have a natural affinity for metal surfaces and may form a protective layer over the surface and reduce the deposition rate. However, they may also react vigorously on the surface, increasing the deposition rate. Heteroatoms of sulfur and nitrogen might also be sources of molecular polarity, and oxygenated hydrocarbons created during fuel storage might also participate in deposit formation reactions; examples of such reactions are suggested in [3.61] and [3.62]. Different fuels produce deposits of similar molecular size distribution and chemical structure, suggesting that there are a limited number of reactive species and that most of these are produced to some extent by most fuels [3.62].

It has been argued [3.17, 3.45] that IVDs are mainly derived from engine oil based on the fact that the concentration of metals such as zinc and calcium, which generally come from engine oil, are 10–30 times higher in the deposits than in the original oil. However, even higher concentrations of metal can be found in the deposits if the fuel contains metal additives. For instance, lead concentrations in IVDs can be over 500 times higher than in the fuel [3.42]. In engines running normally on practical gasolines, IVDs are derived from the fuel with a contribution from the lubricant. In PFI engines, where the fuel is injected on the back of the valve, the fuel plays an important role in controlling IVDs by washing away deposit precursors. In general, IVD levels are lower in PFI engines than in DISI engines or carburetor engines, where such washing action is not available. The engine lubricant seems to play a greater role in IVD formation in DISI engines compared to PFI engines [3.63, 3.64]. IVDs in DISI contain 10% by weight or higher noncarbon (inorganic) elements, of which Ca, Mo, Zn, P, and S, presumably originating from the lubricant, are dominant. The concentrations of these inorganic elements in DISI engine IVDs is at least an order of magnitude higher than in IVDs from PFI engines [3.64].

3.1.3 Combustion Chamber Deposits

Two good sources of information on combustion chamber deposits (CCDs) are [3.65] and [3.66]. The literature before 1995 is reviewed in [3.40].

3.1.3.1 Growth of Combustion Chamber Deposits

CCDs are derived primarily from the fuel, and to some extent from the engine lubricating oil, in typical modern engines with low oil consumption running on typical, full-boiling-range fuels. Deposit formation appears to start with radical-initiated addition or substitution reactions on fuel and lubricant components to produce coordinated oxidation products; these products then condense on the hot surfaces and undergo polymerization [3.67, 3.68]. Lauer and Friel [3.69] suggested an "oxidative pyrolysis" of deposit precursors that condense on the surfaces. This process starts with the reactions caused by radicals that diffuse into the quench layer next to the wall from the flame [3.68]. The condensation of the deposit precursors on the surface is the critical step in this process, and deposit formation is strongly dependent on the surface temperature, given that higher surface temperatures reduce deposit levels [3.70–3.74].

As the deposits grow, the surface temperature increases, because of the insulating properties of the deposits [3.70, 3.75, 3.76]. Hence, as CCDs grow, only the less volatile deposit precursors will condense on the surface and contribute to further deposit growth. Eventually, the surface temperature will be too high for any deposit precursor either from the engine oil or the fuel to condense, and so deposit growth stabilizes. This mechanism is confirmed by the composition variation across the thickness of the deposit with high concentrations of low-melting-point, carbon-rich materials near the wall and high concentrations of less volatile inorganic compounds of calcium and zinc near the deposit surface [3.69, 3.77–3.79]. For a given fuel, there is a maximum surface temperature above which no deposit will form. This critical temperature is around 310 °C for gasoline and about 60 °C higher for engine oil [3.71, 3.80]. For low-boiling-point fuels like iso-octane [3.69, 3.73, 3.74], methanol [3.81], alkylate [3.82], and hydrogen [3.83], this critical temperature will be lower than any normally found in the combustion chamber, and very little deposit from the fuel will be formed. In such cases the deposits will primarily be derived from engine oil. For a given engine operating condition and, hence, a surface temperature regime, if higher-boiling-point materials, such as some fuel additives, are added to a given fuel, more deposits will be formed [e.g., 3.71, 3.72, 3.75, 3.84]. The cleaner the original fuel is, the more significant the increase in deposits caused by the additive is likely to be [3.85].

As the deposits are being formed, they are also being removed through various mechanisms [3.86], including chemical mechanisms like oxidation and gasification [3.87], physical mechanisms like desorption and evaporation of volatile and gaseous components and mechanical removal like abrasion, flaking brought about by thermal stresses [3.77, 3.88], and mechanical wash-off. With many of these mechanisms (e.g., oxidation), as the surface temperature increases, the rate of deposit removal increases [3.87]. Deposit weight and any other property measured at a given time reflect the balance of the deposit formation and removal processes up to that time. CCD levels in an engine can be reduced to a new equilibrium level if some deposit formation is switched off and deposit removal rate is increased, say, by increasing the surface temperature for a given fuel [3.80]. Similarly, for the same engine condition, replacing the fuel with a more volatile fuel, such as an iso-octane and n-heptane mixture [3.89], will switch the deposit formation off temporarily because the surface temperature will be above the critical temperature for the new fuel and deposit weights, and thicknesses will move to a new, lower equilibrium. In general, surface temperatures increase with engine speed, throttle opening, coolant temperature, and, to a lesser extent, with ignition advance and compression ratio except for exhaust valve temperatures, which decrease slightly with an increase in compression ratio [3.19, 3.20]. They also increase as the mixture is made richer, because of the increase in the temperature of the burnt gas, peaking at a fuel/air ratio about 15% fuel rich compared to stoichiometric conditions [3.20]. Hence, increasing the engine speed or load [3.74, 3.90], coolant temperature [3.72, 3.80], and the equivalence ratio of fuel/air mixtures [3.80] above the stoichiometric level leads to lower CCDs. In fact, CCD levels can be substantially reduced in a dirty engine by running it briefly at high load [3.77, 3.91, 3.92].

The conclusions from the pioneering study by Shore and Ockert on the effect of fuel properties on CCDs [3.73] have been confirmed in many other studies. The most important fuel property is the boiling point, and the most critical fuel components in deposit formation are those with the highest boiling points [3.93]. Considering the major gasoline components, at a given boiling point, aromatics are the most prone, and paraffins are the least prone, to form CCDs; olefins are in between the two for CCD formation. Boiling point is the only factor determining the deposit-forming tendencies of aromatics [3.74]. Cyclic alkanes and olefins show greater deposit-forming tendencies compared to their straight-chain counterparts [3.68]. Polynuclear aromatics (PNAs) are strong deposit formers depending on their structure; angular arrangements of aromatic rings produce more deposits than do structures with linear configuration of rings [3.68]. Also, the deposit-forming tendencies of PNAs increase with the number and length of the alkyl side chains attached to the backbone of the molecule [3.68]. Many empirical studies have been published showing that the CCD formation tendency of a fuel increases as its aromatic content increases [3.67, 3.94–3.98]. However, in all these studies the effects of changing composition are not decoupled from the effects of changes in the volatility characteristics. It is possible that the observed changes in CCD formation in many of these reported tests result primarily from changes in the boiling characteristics of the fuel. A parameter derived from both the boiling point and the degree of unsaturation present in a precursor hydrocarbon—termed the *nominal boiling point*—correlates well with deposit-formation rates in a laboratory rig spanning over 4 orders of magnitude with different hydrocarbon molecules [3.68].

Modern detergent packages, many of which contain carrier oils and other co-additives and have high boiling points compared to the fuel, can lead to an increase in CCD formation in unleaded gasolines depending on their concentration and composition [3.40, 3.97]. In general, ASTM unwashed gum (UWG; Section 2.2.1.4) increases in the presence of the detergent additive packages that are usually present in modern gasolines. However, in fuels without such additive packages, UWG does not predict CCD. It is not possible to use UWG or any aspect of its behavior in a thermogravimetric analysis (TGA) test to assess the CCD-forming tendencies of randomly chosen fuels [3.99].

3.1.3.2 Composition, Properties, and Structure of Combustion Chamber Deposits

During an engine cycle, surface temperature can increase by 400 °C from its baseline value in the cooler areas [3.87]. Average piston surface temperatures vary between 200 and 300 °C depending on location and engine operating condition [3.19, 3.20]. Piston temperatures are also 10–25 °C higher than corresponding combustion chamber surface temperatures [3.19]. The exposure of the surface to liquid fuel or additive droplets or to combustion products that might be important in deposit removal and formation mechanisms will also not be uniform across the cylinder. Hence, there are substantial differences in the weight, thickness, and properties of deposits from different parts of the combustion chamber [3.72, 3.100–3.102]. For instance, deposits on exhaust valves, where surface temperatures are high, are composed almost entirely of inorganic compounds [3.73, 3.78, 3.81, 3.100]. Deposits in the end-gas region (i.e., the region where the flame front arrives last) have lower oxygen content and higher carbon content [3.81, 3.100], indicating a less oxidized state, compared to deposits from other parts. End-gas deposits also have lower quantities of volatile material compared to nonvolatiles, which can be interpreted as indicating condensation of heavier aromatics in this area [3.81]. Substantial differences can also be present (e.g., in hydrogen/carbon [H/C] values) in deposits from different engines [3.81, 3.100] and the same engine operating at different conditions [3.100].

However, in general, apart from deposits on exhaust valves, in comparison with fuel and oil, CCD from unleaded gasolines have much higher weight concentrations of oxygen (16–25%) and nitrogen (1–2.5%) and much lower H/C ratios (0.6:1 compared to 1.7:1.9 molar) [3.72, 3.79, 3.100, 3.102, 3.103]; the carbon content is between 60% and 70% by weight. Thus, a big change of the fuel and oil involving a large loss of alkyl components and a significant degree of oxidation take place before their inclusion in CCD. The deposits also have a substantial volatile component; around 35% by weight of the initial deposit is lost if it is heated to 500 °C [3.81, 3.102]. The presence of lead increased deposit weights considerably—by a factor of up to 8 over unleaded gasoline. The weight of the deposit was not related to lead concentration in the fuel above a limit of about 0.15 g/l [3.1]. In deposits from leaded gasolines, lead constituted a substantial (up to 70% by weight) proportion of elemental composition [3.79]. Engine lubricant also contributes to CCD formation [3.104, 3.105].

Densities of around 1530 kg/m^3 have been reported for deposits removed from the combustion chamber for commercial unleaded fuels [3.104]. In other studies where deposit volume was estimated through detailed thickness measurements, densities of 1100–2000 kg/m3 were found with unleaded gasolines [3.40]. Measured heat capacities of deposits removed from the combustion chamber vary from 0.84 to 1.84 kJ/kgK [3.78, 3.104]; there appears to be a good positive correlation between the carbon content of the deposit and its heat capacity [3.78]. Thermal conductivity is more difficult to measure and, for deposits taken out of the engine, depends very much on the way they were prepared into pellets prior to measurement [3.78]. These values have ranged from 0.17 to 0.8 W/mK (see Table I in [3.106]). Thermal characteristics are likely to depend on the morphology of the deposit and need to be measured in situ. Such methods have been developed and depend on making in situ temperature measurements and inferring the thermal properties by modeling the heat transfer through the deposit [3.75, 3.76, 3.88]. Thermal diffusivity values around $0.85 - 4.2 \times 10^7$ m^2 s^{-1} were measured in [3.75, 3.88]. A similarly wide variation in the thermal diffusivity from 0.59×10^{-7} to 3.4×10^{-7} m^2/s in deposits from different locations in the engine has been reported [3.107]. Thermal conductivity and diffusivity of CCD do not depend on engine speed and load [3.108].

The most commonly measured deposit properties are thickness and weight [e.g., 3.66, 3.82, 3.89, 3.101, 3.109]. Such measurements show the differences in deposit thickness in different parts of the combustion chamber (e.g., lower thicknesses in hotter regions) [3.70, 3.75, 3.109]. Stabilized local thicknesses of up to 160 microns are reported in [3.85]. Piston-top deposit weights are usually around 45% of the total CCD weight [3.40]. In a given engine test with a fixed fuel/additive combination, there is usually a correlation between any of the thicknesses measured and deposit weight, although there is a lot of scatter [3.82, 3.101].

Metallic elements, predominantly from the oil, also show up in the deposits; though, naturally, they are more prominent (2–6%) in deposits from unleaded gasolines [3.79, 3.102]. Nevertheless, the dominant oil component for CCD formation is the base oil—increasing the high-molecular-weight and low-volatility content of the oil increases CCD formation [3.110–3.112]. The concentrations of Zn, Ca, and P, all coming from lubricant additives, are up to four times higher in piston-top deposits compared to head deposits [3.72, 3.104, 3.105]; thus engine oil appears to affect the formation of piston deposits more as compared to head deposits [3.72, 3.105].

Different descriptions of the CCD structure have been given from different viewpoints. The carbon is amorphous (nongraphitic), extremely porous, and characterized by a heterogeneous granular structure representative of a carbon produced by a pyrolytic mechanism [3.104]. The exhaustive analytical studies of deposits from unleaded fuel by Ebert and coworkers [3.102] show that CCDs are volatile carbonaceous materials more like bituminous coal than graphite. Up to 50% of the deposits formed inside a DISI engine are reported to consist of volatiles, and the concentration of the volatiles decreased in the following order: cylinder head, piston squish surface, and piston bowl [3.113].

Between 85 and 90% of the carbon is aromatic, and this aromaticity is higher than that found for coals of comparable chemical composition. Their structure is a composite one consisting of a refractory "skeleton" that is rich in oxygen and poor in hydrogen compared to the volatile, smaller molecules adsorbed on this backbone [3.102]. Chapman and Williamson [3.79] reported on columnar and dendritic microstructures made up of inorganic material with amorphous carbon filling the gaps between these structures. The volatile material, which is not bound so tightly to the deposit, closely resembles the heavy ends of the fuel from which the deposit is probably formed. Unburned fuel components and oxidation products derived from fuel and oil can be detected in the deposits; these include carboxylic acids, metal carboxylates, esters, ketones, aldehydes, and lactones [3.94].

The primary molecular building blocks appear to be 2-4 ring PNAs originating in the fuel [3.67]. These building blocks appeared to be independent of the fuel composition or the severity of the test cycle in a given production engine [3.67]. CCD structure is such that they can adsorb and desorb fuel components and combustion products during the engine cycle [3.114]. High-aromatic fuels have been shown to yield condensed, high-aromatic deposits, and low-aromatic fuels produce "fluffier" deposits [3.67, 3.95]. The structures of the deposits on the head and on the piston were not found to be substantially different, but there were cylinder-to-cylinder variations probably caused by temperature gradients [3.67]. The molecular structure of deposits from two production engines running on two different cycles was similar [3.67]. The broad chemical composition and structure of CCDs in DISI engines are similar to that found in PFI engines [3.64, 3.115] and does not depend very much on whether the fuel contains an oxygenate [3.113] or is a synthetic fuel [3.115].

3.2 Effects of Engine Deposits on Engine Performance and Emissions

There have been many studies on the effects of deposits on engine performance and emissions [3.116–3.171]. The tests used to study deposit effects are usually complex, expensive, and marked by poor repeatability and reproducibility. There are also large variations in deposit effects among individual cars when well-controlled fleet tests are conducted. It is also difficult to disentangle the effects of injector deposits, inlet system deposits, and CCDs. The mechanisms by which deposits affect engine performance or emissions cannot be quantitatively understood because of the complexities of the processes involved. However, after many decades of research, most of the deposit effects can be explained qualitatively.

3.2.1 Injector Deposit Effects

Significant reductions in power and driveability can be caused by the restriction of the fuel flow by PFI deposits [3.2, 3.10, 3.15, 3.23]. DISI injector fouling has been linked to statistically significant increases in regulated emissions, particulates, fuel consumption, and poorer driveability [3.116]. Increases in HC and CO emissions have been linked to a disruption of the spray, leading to increased wall film formation [3.117] in an engine different from the one considered in [3.116]. Increased emissions have also been reported with PFI deposits and have been attributed to the inability of the engine control system to correct the imbalance in the mixture strength caused by different levels of fouling among different cylinders [3.23]. Diesel injector fouling and the associated reduction in fuel flow rates will lead to a reduction in power, and internal deposits that cause needle stick will lead to driveability problems and higher emissions. In general, disruption of the injection event and a change in the fuel spray (which is very important for getting the right mixture preparation in DISC as well as modern diesel engines) can be expected to lead to some problems.

3.2.2 Intake Valve Deposits Effects

IVDs could cause power losses [3.10, 3.23, 3.118, 3.119] by reducing the airflow to the engine at full throttle, which in turn could lead to an increase in acceleration times. They can also disrupt the airflow pattern and have a deleterious effect on all aspects of combustion in engines that rely on getting the right mixture preparation in the cylinder [3.10, 3.118]. In PFI engines, IVDs are known to cause driveability problems such as hesitation and poor throttle response during cold start and warm-up [3.47, 3.52, 3.54, 3.120]. IVDs can disrupt the fuel/air mixture formation in such a way as to cause excessively lean mixtures by absorbing unvaporized fuel [3.47]. Recent measurements demonstrate the porosity and the ability of IVDs to absorb fuel components and thus affect mixture preparation [3.121]. Several well-controlled tests have demonstrated around 1% benefit in fuel economy of controlling deposits through an additive package in PFI engines in fleet tests [3.168, 3.122] and in an engine test [3.124]. The road tests also show statistically significant benefits in emissions (e.g., a reduction in HC of around 11%) [3.168]. These additive packages also keep the injectors clean and might affect CCDs, so special efforts are required to isolate the effects of IVDs [3.124]. In fleet tests, this is achieved by comparing two well-matched vehicle fleets, one running on base fuel and the other on the additive package. IVDs have also been shown to increase NOx emissions [3.10, 3.118] at part-throttle operation. Single-cylinder engine tests [3.118] suggested that this was because the combustion rate was increased, presumably by IVDs heating the intake air by transferring stored heat. In some engines, IVDs have been shown to contribute to octane requirement increase (ORI) [3.48, 3.49, 3.140]; see also Section 3.2.3.1. Again, this might be caused by the heating of the intake charge by IVDs.

3.2.3 Combustion Chamber Deposit Effects

Of all the deposits, the effects of CCDs on engine performance and emissions have received the most attention. The effect of CCDs on heat transfer at the combustion chamber walls has been demonstrated experimentally in many independent studies [e.g., 3.19, 3.75–3.78, 3.92, 3.104, 3.106, 3.107, 3.109, 3.125]. They act as insulators and as heat reservoirs, storing heat in one cycle and releasing it to the fresh charge in the next cycle. They also occupy volume, increasing the compression ratio. Finally, they might also absorb and release unburned fuel and proknock species and promote chemical reactions through catalytic effects. As deposits build up in the combustion chamber, the maximum [3.19] as well as the average [3.109] heat flux away from the combustion chamber decreases, heat lost to the coolant decreases [3.77, 3.126, 3.127], air consumption rate is reduced [3.77, 3.127, 3.128], volumetric efficiency decreases [3.92, 3.128–3.130], and flame propagation rates increase [3.109, 3.126, 3.128]. Thus, the incoming charge is heated by the deposits; Harrow and Orman estimated that deposits increased the temperature of the trapped charge by about 11 °C in their experiments [3.128]. Modeling of the heat transfer through the deposit layer also shows that the deposit surface temperature, the bulk gas temperature, and the end-gas temperature all rise as deposits build up. [3.130–3.135]. Studzinski et al. [3.135] estimated that the bulk gas temperature increases by about 10 °C because of volumetric effects and by about 50 °C from heat transfer changes owing to the deposits.

The observed effects of CCDs on engine operation discussed in the next section can be understood to an extent in the context of the preceding discussion.

3.2.3.1 Octane Requirement Increase

The tendency of a car to knock and, hence, its octane requirement (OR) increase with use as deposits build up. In bench engine tests or vehicle dynamometer tests where operating conditions are either constant or consist of repetitive cycles, OR increases rapidly in the early stages and reaches an equilibrium value as deposits stabilize. However, in field tests, an equilibrium octane number and the distance when equilibrium is reached are sometimes difficult, if not impossible, to define [3.48, 3.105]. It has been argued that concepts such as ORI can be applied to car populations only in statistical terms [3.48].

Typical values of ORI ("stabilized" OR minus initial OR) are between 3 and 5 for modern cars running on unleaded gasoline over 12,000 miles, and there may be large variations between individual cars of the same model, from 0 to 15 ORI. New cars, even of the same model, can have widely varying octane requirements, but with use, the octane requirements are distributed in a narrower band, so that the larger the initial octane requirement (IOR), the lower the ORI [3.48, 3.78, 3.83,3.137]; there is an empirical correlation, given by ORI = A – B(IOR), where A and B are constants. The value of B was around 0.4 in both U.S. [3.48] and Japanese [3.78] cars.

In very general terms, octane requirement is increased by factors that tend to increase combustion chamber deposit quantity, such as fuel properties like high boiling points [3.137], high aromatic content [3.50, 3.96, 3.97], and the presence of some additives [3.111, 3.136, 3.137]. However, some fuel additive packages have been shown to be either neutral or beneficial (i.e., reducing ORI), compared to base fuel [3.133, 3.139] in some engine and road tests. Similarly, factors that cause the removal of CCDs, such as hard driving modes [3.77, 3.83, 3.91, 3.111, 3.137] or running a dirty engine on iso-octane [3.127], can reduce the octane requirement. In well-controlled engine tests a reasonable overall correlation between CCD weight and ORI can be established [3.78]. However, this is the exception rather than the rule; usually, CCD weight does not correlate with ORI [3.66]. Deposits in different parts of the combustion chamber have varying degrees of effect on ORI; this has been studied by selective sequential removal of different deposits. For instance, deposits in the end-gas region (i.e., the region where combustion takes place last) appear to be the most significant [3.111], whereas piston-top deposits contribute much less to ORI compared to head deposits [3.109]. In some engines inlet valve deposits contribute significantly to ORI [3.48, 3.49, 3.133, 3.139, 3.140].

CCDs cause ORI primarily because they increase the mixture temperature by storing up heat from one cycle and giving it up to the fresh charge in the succeeding cycle; the increase in surface temperature caused by CCDs also plays a role. The volumetric effect of CCDs, because they occupy space and thus reduce the compression ratio, accounts for about 10% of their total effect on ORI in unleaded gasolines [3.111]. In addition, there might be a chemical mechanism; one possibility is that CCD absorb stable proknock species like hydrogen peroxide and release them in the end-gas to promote knock [3.89, 3.134]. The extent of such chemical mechanisms and the balance between chemical and physical mechanisms will almost certainly depend on the nature of the deposits and the fuel being used, but it is not understood at this stage. How inlet system deposits affect ORI is also not understood. Thus ORI is a complex phenomenon that is not fully understood even after decades of extensive study.

Testing for ORI, like in other deposit-related studies, is characterized by poor repeatability and reproducibility; it is also labor-intensive and expensive. The difficulties in ORI testing and in drawing sound and useful conclusions from such testing have been highlighted by many studies [e.g., 3.66, 3.105, 3.111, 3.36]. It is important to be aware of such difficulties while assessing the differences between, say, the effect of two additives on ORI.

3.2.3.2 Fuel Economy

Nakamura, Yonekawa, and Okamoto found, in 11 different cars, that fuel economy improved as deposits grew and octane requirement increased—by 13% in the most extreme case. It was confirmed, by removing CCDs and remeasuring fuel consumption [3.78, 3.125], that most of this improvement was caused by CCDs. These cars were running on unleaded gasoline on the road, and fuel economy was measured in a "10-mode" cycle. They also showed [3.78] that in another car that ran on a chassis dynamometer at a constant speed of 60 km/h, fuel economy improved by about 9% in 10,000 km, with most of the change occurring in the first 5000 km. In another experiment [3.78], they coated the cylinder head surface of a research engine with Teflon® and found that the octane requirement increased and fuel economy improved, depending on the thickness of the coating; a 15.4% improvement in fuel economy at a fixed speed and load with a coating 0.08 mm thick were measured.

Kalghatgi et al. [3.109] measured a reduction in specific fuel consumption of about 7% at part throttle, with the reduction being most pronounced at the start of the test, as deposits built up in an engine test. By removing the deposits in stages, they established that almost all this effect was attributable to head deposits, and piston-top deposits had little effect. Spink et al. [3.122] reported that, in three test fleets each with 15 cars and each fleet running on a different unleaded fuel and additive combination, on-the-road fuel economy improved by around 17% over 10,000 miles (16,000 km) compared to the clean condition at the start of the test. Almost all of this improvement, part of which can be attributed to the reduction in friction losses during "running in," was in the first 3000 miles (4800 km).

The fuel economy improvement for a given fleet was 2.4–4.5% in an ECE test cycle [3.122]; the initial measurement in this case was 1000 miles (1600 km) after the start of the test and the final measurement, after 10,000 miles (16,000 km). Fuel economy appears to change rapidly at the start of deposit buildup. If the "clean" engine measurement is not made in a truly clean engine, the first measurement of fuel consumption and the apparent change brought about by the deposits will be smaller. The fuel economy improvements reported by Spink et al. [3.122] cannot be attributed to CCDs unequivocally because they did not make any "check-back" measurements by removing CCDs.

Graiff measured fuel consumption at different speeds and road load conditions using two bench engines and a vehicle [3.139]; removing CCDs caused an increase of about 2% in fuel consumption at all conditions. A similar figure is quoted by Woodyard in [3.66] for two bench engines. Tsutsumi et al. [3.140] measured fuel consumption at wide-open throttle at different speeds in an engine test as part of their investigations on the effects of the surface finish of the combustion chamber. Their Fig. 15 shows that when CCDs built up, brake-specific fuel consumption decreased by about 2% at all speeds up to 6000 rpm. Yonekawa et al. assessed the changes caused by CCDs on the energy balance in a single-cylinder engine by analyzing the pressure curves and attributed the improvement in fuel economy at a fixed speed and load primarily to a reduction in heat lost to the coolant [3.126]; faster flame development also plays a role. The heat lost to coolant constitutes a much larger fraction of fuel energy at low speeds and loads than at full throttle operation [3.141]. Therefore, the fuel economy benefits of CCDs are likely to be less marked at high loads.

Thus, CCDs generally have a beneficial effect on fuel economy for the same reasons that they increase the octane requirement, though the extent of this improvement might be different in different engines and operating conditions.

3.2.3.3 Combustion Chamber Deposit Flaking and Associated Problems

Small flakes of CCDs can break loose and get trapped between the exhaust valve and the seat. This can result in difficulties in starting, and increase in hydrocarbon emissions, and rough running [3.142–3.145]. Serious field problems associated with this phenomenon in the past have been reported in the United States and Europe. The problem usually occurs in cars that have run for 5000–10,000 miles (8000–16,000 km) on mild, usually urban-driving, cycles. Typically, the owner moves the car a short distance, say, from in front of the house to the garage at night when the engine is cold. The next morning the car suffers from starting problems. Incidence of the problems peaked during the winter months in Europe when they occurred. It has been established that this problem occurs because of flaking of CCDs during the period when the car has been stationary. These flakes get trapped between the exhaust valve and its seat when the engine is started and eventually will cause a loss of compression in the cylinder. Deposit flaking could be directly connected to increases in leakage through the valves and startability problems in three car models [3.145]. The nonstart problem is an extreme manifestation of the deterioration of exhaust valve seating caused by deposit flakes. It is possible that this problem is more common than realized and is not usually detected. Some engine professionals have suggested in private communications that onboard diagnostic indications of malfunctions associated with increased emissions in modern engines are often due to deposit flaking.

The flaking most likely occurs because of the interaction of water with CCDs. During normal operation, water vapor from the atmosphere or from combustion before the engine was last switched off can condense on the deposits when the engine is standing and provide the trigger for deposit flaking. This is obviously far more likely to happen when the engine is cold and remains cold after a very brief period of running, as is likely in winter. Figure 3.7, from [3.144], shows an example of CCD flaking brought about by water. Figure 3.7a shows the piston top covered in deposits from a six-cylinder engine after it has been run for 3000 miles (4800 km). Figure 3.7b shows the same piston top two days after it has been lightly sprayed with water; some of the deposits have flaked. These flakes are loose and can be easily removed by gently blowing on them, as shown in Fig. 3.7c. Spraying fuel onto deposits can cause them to flake as well, but water is much more effective in causing flaking [3.143, 3.144].

In many cases [3.144] significant flaking can occur even without spraying the deposits with water; absorption of ambient water vapor by the deposits in a car left standing for a few days seems sufficient to cause flaking. In some cases, upon placing a drop of water on the deposit, the drop disappears within a few minutes, the deposit surface blisters where the drop had been placed, and the flakes can come loose with sufficient violence to jump several centimeters off the metal surface [3.143]. The thermal regime under which deposits are formed is important in determining whether the deposits flake or not [3.143, 3.144]. In road tests a mild driving cycle produced deposits that were more prone to flaking than a hard driving cycle [3.144]. Detergent additive packages used to control injector and IVDs generally suppress deposit flaking [3.143, 3.144], presumably by making the CCDs less permeable to water.

3.2.3.4 Emissions

NOx emissions are reduced when CCDs are completely removed [3.135, 3.146–3.149]. In fleet tests, NOx generally increases as the test progresses and deposits build up [3.122, 3.150]. The results of Studzinski et al. [3.135] suggest that piston-top deposits have a smaller effect compared to head deposits on NOx; their modeling study relates the increase in NOx to the increase in bulk gas temperature brought about by CCDs. In a study using a DISI engine,

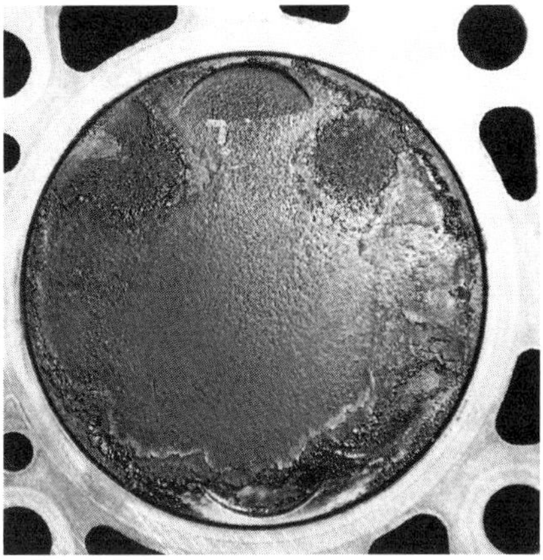

(a)

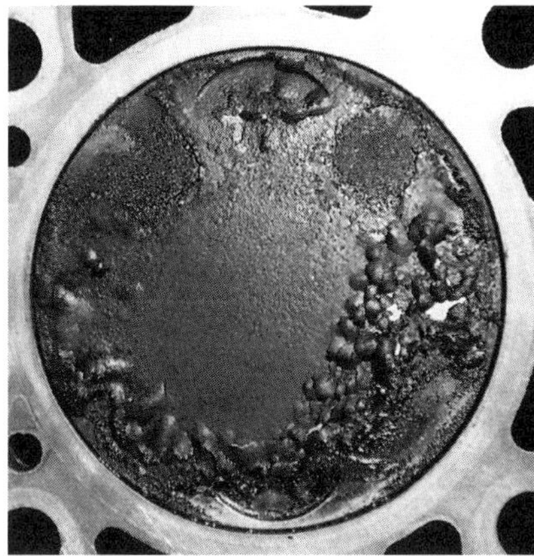

(b)

(c)

Fig. 3.7. Combustion chamber deposit flaking [3.144]. (a) Piston-top deposits from one of the cylinders from a six-cylinder engine run for 3000 miles (4800 km) on the road. (b) Deposits shown in Fig. 3.7a two days after spraying with water. (c) Deposit flakes shown in Fig. 3.7b can be removed by blowing lightly on them. (Copyright by Shell Global Solutions. Reproduced with permission.)

Watanabe et al. [3.151] reported that CCDs increased NOx emissions under low-speed stratified charge operation.

There are at least three mechanisms through which CCDs can affect HC emissions. First, an increase in combustion chamber surface temperatures brought about by increasing the coolant temperature, for instance, reduces HC emissions [3.149, 3.152–3.154], and since CCDs increase surface temperatures, they might be expected to reduce HC emissions. Second, it is possible that deposits in the crevices could reduce crevice volumes, which are major sources of HC emissions; CCDs could reduce HC emissions through this mechanism [3.149]. Third, CCDs can absorb fuel HCs and release them, largely unchanged, during the exhaust stroke [3.82, 3.149, 3.155, 3.156]. Presumably, depending on which of these mechanisms predominate, CCDs cause either an increase [3.147, 3.149, 3.156–3.158] or a decrease in HC, as two road tests have shown [3.122, 3.150];

in these fleet tests, IVDs and fuel system deposits were controlled using additives. Two other studies [3.146, 3.159] reported no statistically significant effect of CCDs on HC.

The empirical evidence for the effect of CCDs on CO is similarly mixed [3.40]. CO_2 levels decreased in the fleet tests reported in [3.122, 3.150], consistent with the observation that fuel consumption decreased, as discussed in Section 3.2.3.2.

3.2.3.5 Combustion Chamber Deposit Interference or Carbon Rap

Most modern engines have flat areas around the rim of the piston that cause radial gas motion known as "squish" as they approach similar flat areas on the head during the compression stroke. At top dead center, the clearance between the head and the piston will be the smallest in the squish area. Reduce squish clearances, to as low as 0.7 mm, has become a design trend to help reduce emissions [3.160]. The squish clearance will inevitably be less than the design value in some cars because of production tolerances. In addition, when the engine is cold, the piston is tilted slightly as it rises [3.160], particularly at low engine speeds [3.72]. This rocking motion further reduces the minimum clearance between the piston and head. This minimum clearance can be eliminated by CCDs, and the piston top can actually hit the head surface under some circumstances. The clattering sound that results in the frequency range of 1–10 kHz [3.72, 3.82, 3.161] has been variously described as carbon rap, carbon knock, deposit-induced noise, and combustion chamber deposit interference (CCDI). The problem was experienced by several engine manufacturers and drew renewed attention to CCDs in the early 1990s, leading to a special workshop on CCDs in 1993 that discussed the problem extensively [3.66]. However, it seems to have been cured by design changes to engines.

3.2.3.6 Other Effects of Combustion Chamber Deposits

The maximum engine power at full throttle can be reduced by CCDs because they increase intake charge temperatures and cause a reduction in the mass of air that can be breathed in. However, as a dirty engine is run at full throttle, the CCD level can be reduced (see Section 3.1.3.1), and there will be a steady increase in power [3.92]. The final stabilized power can be 6–10% lower than for the clean engine [3.92, 3.129]; these assessments need to be made in modern engines using unleaded fuels. CCDs have also been implicated significantly in preignition (see Chapter 5) in older engines.

Spark plug fouling can be caused by deposits on the body of the plug. Greenshields published an authoritative study on plug fouling caused by deposits from leaded gasolines on the ceramic core of the spark plug [3.161]. An occasional modern problem is plug fouling caused by carbon deposits during the numerous very short trips taken by a car during its production and initial delivery [3.162–3.164]. Moisture deposition on the carbon layer causes a conductive layer between the central electrode and the shell, causing charge leakage and failure of the plug-gap to fire [3.163, 3.164]. In contrast, low-electron work-function deposits produced by spark-aider fuel additives on plug electrodes, particularly the cathode, can improve the energy-transfer efficiency of the spark. This can lead to faster early flame development and a big improvement in the ignition ability of the spark near the ignition limits [3.165–3.167], which results in reduction in misfires, extension of the lean limit and improvements in driveability [3.169]. However, these additives contain low work-function metals like potassium and are no longer used because of concerns about their effects on catalysts.

Deposits that form in the ring belt area could cause piston-ring sticking or ring plugging and ring breakage, all of which could lead to engine malfunction and short engine life [3.171]. Such deposits are likely to be more associated with the lubrication system than the fuel/combustion system. Piston deposits are evaluated in standard test methods like ASTM D7320, which is used to evaluate engine oils for SI engines, but these deposits are not in the combustion chamber and do not directly affect combustion.

Table 3.1 lists the major factors affecting deposit formation in engines, and Table 3.2 lists the effects of deposits on engine performance and emissions.

3.3 Deposit Control

The design of the engine and the engine components affects deposit formation and the effects that deposits have on engine operation. For instance, the positioning and design of the port fuel injectors, as well as their surroundings, will affect the average tip temperatures during a hot soak and will affect PFI deposit formation [3.23, 3.25]. EGR and PCV flows, required for emissions control, will also affect deposit formation. Sometimes engine features can be redesigned to avoid excessive deposit formation or to alleviate the effects of deposits. For example, early DISI engines suffered from very high levels of IVDs. This problem appears to have been alleviated by redesigning EGR flows and oil flow down the valve stem. Carbon rap seems to have been designed out by simply increasing squish clearances. Problems associated with CCDs flaking also appear to have been cleared up because the field problems seem to have disappeared without any specific changes made to the fuels or additives. It is not clear how this has been accomplished, but it might have included redesign of the injection system and the strategy to change the interaction of liquid fuel with CCDs.

The most common approach to deposit control is through the fuel and, to some extent, lubricant composition. However, it is very difficult and not economical

Table 3.1 Major Factors Affecting Deposit Formation (with an increase in the parameter in the first column)

	Injector Deposits				
	PFI SI Engines	DISI Engines	Diesel Engines	IVDs in PFI Engines	CCD in SI Engines
Fuel olefin content	Increase			Increase	
Fuel aromatic content					Increase
Fuel volatility		Increase (e.g., as T90 decreases)		Increase	Decrease
Detergent additives	Decrease	Decrease	Decrease	Decrease	Increase with some additives
Inlet temperature				Increase	
Surface temperature	Increase	Increase	Increase	Increase up to a limit	Decrease
Oil flow down the valve stem				Increase at low rates and then decrease	Increase

to control the composition of base fuel purely with deposit control in mind. In the past, imperatives such as increasing the antiknock quality of the fuel or refinery yield tended to increase olefin levels, which were sometimes bad for IVDs and PFI deposits, in U.S. regular grade gasoline [3.15]. Moreover, in most markets, fuels retailers exchange products, so they do not have full control on the base fuel composition. Therefore, the use of fuel additives is the most practical option for controlling deposits in PFI [e.g., 3.10, 3.15, 3.122, 3.123, 3.170, 3.172], DISI [e.g., 3.116, 3.173, 3.174] and diesel [e.g., 3.31, 3.32] injectors, and IVDs in PFI engines [e.g., 3.176, 3.23, 3.122, 3.123].

Typical additive dose rates are 30–200 ppm of active material—over half of an additive package can be made up of a solvent. The lower, "keep-clean" dose rates are meant to keep deposits from building up in clean engines, while at the higher, cleanup dose rates, additives can remove deposits that have already built up. In PFI engines, a modern additive package used at a dose rate required to keep IVDs at acceptable levels can also keep the injectors clean. Note that there is no practical additive solution to controlling CCDs, though it is reported in [3.175] that a very high (>0.2% active material) dose of

Table 3.2 Effects of Deposits

Operating Parameter	Deposit Effect Compared to a Clean Engine
Fuel economy	Worsens, because of injector deposits and IVDs in PFI engines—road tests demonstrate ~1–2% improvement when injectors and valves are kept clean with additives. However, fuel economy improves because of CCDs in SI engines by up to 15% since heat loss to coolant is reduced.
Maximum power	Decreases, because of the decrease in fuel flow rate caused by all injector deposits and reduction of volumetric efficiency caused by CCDs.
Driveability	Impaired by all injector deposits, because of fuel flow restrictions, needle sticking from internal deposits in diesel injectors, fuel holdup in IVDs in PFI engines, and CCD flakes degrading valve sealing.
Emissions	Increase, because of injector deposits—probably because of disruption of the spray pattern and because of the reasons causing impaired driveability. CCDs increase NOx because of higher combustion temperatures.
Octane requirement (OR)	Increases, because of CCDs, primarily because of the heating of the intake charge. In a few engines IVDs also contribute to OR increase.
Startability	May decrease in SI engines, because of spark plug fouling or loss of compression because of CCD flakes.

an additive based on poly ether amine chemistry reduced CCD levels. In this case the deposit removal mostly occurred in the combustion chamber head around the inlet valve—the additive seemed to display detergency capability as on the intake system. Lubricant formulation has a far bigger effect on IVD control in DISI engines where there is no direct interaction between the fuel and IVDs [3.63].

The main active ingredient in a deposit control additive (DCA) package is a detergent or dispersant. Some typical compounds used for deposit control are amino amides, polyether and polybutene amines, alkyl and polybutene succinimides, and hydroxyl polyamino carbamates. They are surface-active materials made up of a hydrocarbon-soluble polymer connected through a suitable linking group to a polar head [3.172, 3.176]. The selection of the polymer is critically important to minimize undesirable side effects. The polar heads generally contain nitrogen and sometimes oxygen heteroatoms, which are attracted to metal surfaces or deposit precursors, while the hydrocarbon tails are projected into the fuel [3.172]. The molecular weight of the DCA is primarily determined by the length of the hydrocarbon tail. Detergents function by causing a thin hydrocarbon film to be held against the metal surface to prevent deposition or adhesion of deposit precursors on the surface [3.22, 3.177] in the keep-clean mode. DCAs could also react with deposit precursors to form micelles and inhibit deposition [3.22] (i.e., act as dispersants); additives in the lubricant in particular could be acting in this manner [3.42].

It has been suggested that high-molecular-weight (~1000 MW) DCAs are better at dispersion of deposit precursors because their long hydrocarbon tails allow them to be more easily dissolved by the fuel, whereas low-molecular-weight (<500 MW) DCAs are better at adsorption on the metal surface [3.172]. At the higher dose rates used in the cleanup mode, a DCA is believed to act by dissolving the soluble part of the deposit, which probably binds the deposit to the surface [3.22, 3.177]. In this mode, the additive moves with the deposit and must be replaced at a faster rate and, hence, needs to be used at a higher concentration compared to the keep-clean mode [3.177]. DCAs are sometimes used with carrier fluids that are high-boiling-point, thermally stable solvents [3.49, 3.51, 3.58]. An analogy is soap and water—the carrier is the water [3.170]. Carriers are capable of dissolving some of the deposits and, by themselves, control some types of deposits [3.170, 3.49].

The effectiveness of a DCA package depends on the nature of deposits, which can be different for different gasolines and different engine operating conditions. The additive package also needs to be tailored to tackle the deposit in a particular area. Thus, low-molecular-weight detergents successfully used in carburetor engines performed poorly in PFI engines, causing IVDs to increase substantially, which prompted the development of new detergent chemistry.

Possible side effects have to be properly assessed and guarded against when a new additive package is introduced into a market [3.170, 3.172, 3.175, 3.176]. Side effects include storage and handling problems such as water shedding, storage stability, additive separation and depletion, filter plugging in storage tanks, emulsion formation, and dirt suspension; material compatibility (e.g., the effects on seals in the fuel system); toxicity; and lubricant oil degradation. The last of these problems arises because of fuel additives getting into the lubricant, usually resulting in a progressive increase in lubricant viscosity [3.175, 3.176]. However, benefits from this process have also been claimed. For instance, DCAs accumulating in the oil can improve engine cleanliness in areas affected primarily by the oil (e.g., sludge control) [3.178, 3.179].

3.4 Other Deposits of Concern in Internal Combustion Engines

In the past, detergent additives sometimes interacted with the lubricant and fuel to produce sticky deposits that formed on the valve stem, which caused the valve to stick open and led to cold start and driveability problems and, in the worst cases, engine damage [3.43, 3.180, 3.181]. EGR coolers are increasingly used in modern diesel engines to cool EGR gases in order to reduce NOx emissions. A highly porous soot deposit layer can form inside the cooler, causing a decrease in its efficiency and an increase in the pressure drop across the cooler. This process can have a big impact on the control of engine-out NOx emissions [3.182, 3.183]. It might be possible to control them by using high surface temperatures with high EGR rates [3.182]. Wetting the deposits with water vapor or condensed water—soot hydration—also makes it easier to remove them [3.183]. Deposits can also form in urea-SCR systems used in modern diesel engines to control NOx.—their formation and control has been studied in [3.184].

3.5 References

3.1. Kalghatgi, G.T. 1990. "Deposits in Gasoline Engines—A Literature Review." SAE Paper No. 902105. SAE International, Warrendale, PA.

3.2 Benson, J.D., and Yaccarino, P. 1986. "Fuel and Additive Effects on Multiport Fuel Injector Deposits." SAE Paper No. 861533. SAE International, Warrendale, PA.

3.3 Lacey, P., Gail, S., Kienz, J-M., Milovanovic, M., and Gris, C. 2011. "Internal Fuel Injector Deposits." SAE Paper No. 2011-01-1925. SAE International, Warrendale, PA.

3.4 Schwab, S.G., Bennett, J.J., Dell, S.J., Gallante, J.M., Kulinowski, A.M., and Miller, K.T. 2010. "Internal Injector Deposits in High-Pressure Common Rail Diesel Engines." SAE Paper No. 2010-01-2242. SAE International, Warrendale, PA.

3.5 Ullmann, J., Geduldig, M., Stutzenberger, H., Caprotti, R., and Balfour, G. 2008. "Investigation into the Formation and Prevention of Internal Diesel Injector Deposits." SAE Paper No. 2008-01-0926. SAE International, Warrendale, PA.

3.6 Quigley, R., Barbour, R., Fahey, E., Arters, D.C., Wetzel, W., and Ray, J. 2011. "A Study of the Internal Diesel Injector Deposit Phenomenon." Paper presented at the Eighth International Colloquium Fuels. TAE, Esslingen, January 19–20.

3.7 Hilden, D.L. 1988. "The Relationship of Gasoline Diolefin Content to Deposits in Multiport Fuel Injectors." SAE Paper No. 881642. SAE International, Warrendale, PA.

3.8 Kim, C., Tseregounis, S.I., and Scruggs, B.E. 1987. "Deposit Formation on a Metal Surface in Oxidized Gasolines." SAE Paper No. 872112. SAE International, Warrendale, PA.

3.9 Tseregounis, S.I. 1988. "Thin Deposit Films from Oxidized Gasoline on Steel Surfaces as Determined by ESCA." SAE Paper No. 881641. SAE International, Warrendale, PA.

3.10 Taniguchi, B.Y., Peyla, R.J., Parsons, G.M., Hoekman, S.K., and Voss, D.A. 1986. "Injector Deposits—the Tip of Intake System Deposit Problems." SAE Paper No. 861534. Also in *Special Publication 713*, "Deposit Formation in Gasoline Fuel Injected Engines." SAE International, Warrendale, PA.

3.11 Caprotti, M., Takaharu, S., and Masahiro, D. 2008. "Impact of Diesel Fuel Additives on Vehicle Performance." SAE Paper No. 2008-01-1600. SAE International, Warrendale, PA.

3.12 Barbour, J., Arters, D., Dietz, J., Macduff, M., Panesar, A., and Quigley, R. 2007. "Diesel Detergent Additive Responses in Modern, High-Speed, Direct-Injection, Light-Duty Engines." SAE Paper No. 2007-01-2001 (JSAE20077144). SAE International, Warrendale, PA.

3.13 Tang, J., Pischinger, S., Lamping, M., Körfer, T., Tatur, M., Tomazic, D. 2009. "Coking Phenomena in Nozzle Orifices of DI-Diesel Engines." SAE Paper No. 2009-01-0837. SAE International, Warrendale, PA.

3.14 Kolobielski, M., and McCaleb, F. 1986. "Gasoline and Engine Oils: Literature Review, New Laboratory Oxidation Method and Significance of Olefins in Fuel." Report No. 2296, U.S. Army MERD Command, Government Accession No. ADA 086740.

3.15 Tupa, R., and Koehler, D.E. 1986. "Gasoline Port Fuel Injectors—Keep Clean/Clean Up with Additives." SAE Paper No. 861536. SAE International, Warrendale, PA.

3.16 Aradi, A.A., Imoehl, B., Avery, N.L., Wells, P.P., and Grosser, R.W. 1999. "The Effect of Fuel Composition and Engine Operating Parameters on Injector Deposits in a High-Pressure Direct Injection Gasoline (DIG) Research Engine." SAE Paper No. 1999-01-3690. SAE International, Warrendale, PA.

3.17 Esaki, Y., Ishiguro, T., Suzuki, N., and Nakada, M. 1990. "Mechanism of Intake-Valve Deposit Formation. Part I: Characterization of Deposits." SAE Paper No. 900151. SAE International, Warrendale, PA.

3.18 Finlay, I.C., Harris, D., Boam, D.J., and Parks, B.I. 1985. "Factors Influencing Combustion Chamber Wall Temperature in a Liquid Cooled Automobile Spark Ignition Engine." *Proceedings of the Institution of Mechanical Engineers* 199(D3):207.

3.19 Hayes, T.K., White, R.A., and Peters, J.E. 1993. "Combustion Chamber Temperature and Instantaneous Local Heat Flux Measurements in a Spark Ignition Engine." SAE Paper No. 930217. SAE International, Warrendale, PA.

3.20 French, C.C.J., and Atkins, K.A. 1973. "Thermal Loading of a Petrol Engine." *Proceedings of the Institution of Mechanical Engineers* 187(49):561–573.

3.21 Greber, A.F., and Smith, R.G., 1971. "Some Effects of Exhaust Gas Recirculation Upon Automotive Engine Intake System Deposits and Crankcase Lubricant Performance." SAE Paper No. 710142. SAE International, Warrendale, PA.

3.22 Marbach, H.W., Johnston, A.A., Bowden, J.N., and LePera, M.E. 1979. "A Novel Laboratory Method for Evaluating Induction System Deposits in Gasoline Engines." SAE Paper No. 790204. SAE International, Warrendale, PA.

3.23 Tupa, R.C. 1987. "Port Fuel Injector Deposits: Causes/Consequences/Cures." SAE Paper No. 872113. SAE International, Warrendale, PA.

3.24 Abramo, G.P., Horowitz, A.M., and Trwella, J.C. 1986. "Port Fuel Injector Cleanliness Studies." SAE Paper No. 861535. Warrendale, PA.

3.25 Cracciolo, F., and Stebar, R.F. 1987. "an Engine Dynamometer Test for Evaluating Port Fuel Injector Plugging." SAE Paper No. 872111. SAE International, Warrendale, PA.

3.26 Lenane, D.L., and Stocky, T.P. 1986. "Gasoline Additives Solve Injector Deposit Problems." SAE Paper No. 861537. SAE International, Warrendale, PA.

3.27 Tupa, R.C., Taniguchi, B.Y., and Benson, J.D. 1989. "A Vehicle Test Technique for Studying Port Fuel Injection Deposits—A Coordinating Research Council Program." SAE Paper No. 890213. SAE International, Warrendale, PA.

3.28 Richardson, C.B., Gyorog, D.A., and Beard, L.K. 1989. "A Laboratory Test for Fuel Injector Studies." SAE Paper No. 892116. SAE International, Warrendale, PA.

3.29 Kinoshita, M., Saito, A., Matsuhita, S., Shibata, H., and Niwa, U. 1999. "A Method for Suppressing Formation of Deposits on Fuel Injector for Direct Injection Gasoline Engine." SAE Paper No. 1999-01-3656. SAE International, Warrendale, PA.

3.30 Ashida, T., Takei, I., Hohi, H. 2001. "Effect of Fuel Properties on SIDI Injector Deposits." SAE Paper No. 2001-01-3694. SAE International, Warrendale, PA.

3.31 Caprotti, R., Bhatti, N., and Balfour, G. 2010. "Deposit Control in Modern Diesel Fuel Injection Systems." SAE Paper No. 2010-01-2250. SAE International, Warrendale, PA.

3.32 Arpaia, A., Catania, A.E., D'Ambrosio, S., Ferrari, A., Luisi, S., and Spessa, E. 2009. "Injector Coking Effects on Engine Performance and Emissions." ASME Paper ICEF 2009-14094. In *ASME (USA), ASME-ICEF*, Lucerne, Switzerland September 20–23.

3.33 Birgel, A., Ladommatos, N., Aleiferis, P., Zulch, S., Milovanovic, N., Lafon, V., Orlovic, A., Lacey, P., and Richards, P. 2008. "Deposit Formation in the Holes of Diesel Injector Nozzles: A Critical Review." SAE Paper No. 2008-01-2383. SAE International, Warrendale, PA.

3.34 Ikemoto, M., Omae, K., Nakai, K., Ueda, R., Kakehashi, N., Sunami, K. 2011. "Injection Nozzle Coking Mechanism in Common-rail Diesel Engine." SAE Paper No. 2011-01-1818. SAE International, Warrendale, PA.

3.35 "CEC F-98-08: Direct Injection, Common Rail Diesel Engine Nozzle Coking Test." Issue 4, March 3, 2011. Coordinating European Council. Desford, UK.

3.36 Williams, R., and Balthasar, F. 2009. "Diesel Fuel Degradation and Contamination in Vehicle Systems." Paper presented at the Fuels 2009—7th International Colloquium, Technische Akademie Esslingen (TAE), Ostfildern, Germany.

3.37 Quigley, R., Barbour, R., and Marshall, G. 2007. "Trace Metal Contamination of Diesel Fuels." Paper presented at the Fuels 2007—6th International Colloquium, Technische Akademie Esslingen (TAE), Ostfildern, Germany.

3.38 Barbour, R., Quigley, R., Browne, D., and Panesar, A. 2011. "A Comparison of Peugeot DW 10 Dynamometer and Vehicle Engine Performance." Paper presented at the Fuels 2011—8th International Colloquium Fuels, Technische Akademie Esslingen (TAE), Ostfildern, Germany.

3.39 Uitz, R., and Brewer, M. 2009. "Impact of FAME Quality on Injector Nozzle Fouling in a Common Rail Diesel Engine." SAE Paper No. 2009-01-2640. SAE International, Warrendale, PA.

3.40 Kalghatgi, G.T. 1995. "Combustion Chamber Deposits in Spark-Ignition Engines—A Literature Review." SAE Paper No. 952443. SAE International, Warrendale, PA.

3.41 CRC. 1979. "Intake Manifold Deposit Engine Dynamometer Test Procedure. State-of-the Art Summary Report. 1973–1978." CR Report No. 505, Coordinating Research Council, Desford, UK.

3.42 Rogers, D.T., and Jonach, F.L. 1958. "Mechanism of Intake Valve Underside Deposit Formation." SAE Paper No. 106C. SAE International, Warrendale, PA.

3.43 Jewitt, C.H., Bostick, G.L., and Kersey, V.L. 1987. "Fuel Injector, Intake Valve and Carburetor Detergency Performance of Gasoline Additives." SAE Paper No. 872114. SAE International, Warrendale, PA.

3.44 Siegel, W.O., and Zinbo, M. 1985. "On the Chemical Composition and Origin of Engine Deposits." In *Chemistry of Engine Combustion Deposits,* edited by L.B. Ebert, Plenum Press, New York, p. 53.

3.45 Nomura, Y., Osawa, K., Ishiguro, T., and Nakada, M. 1990. "Mechanism of Intake Valve Deposit Formation. Part 2: Simulation Tests." SAE Paper No. 900152. SAE International, Warrendale, PA.

3.46 Lepperhoff, G., Schommers, J., Weber, O., and Leonhardt, H. 1987. "Mechanism of Deposit Formation at Inlet Valves." SAE Paper No. 872115. SAE International, Warrendale, PA.

3.47 Tomita, M., Okada, M., Katayama, H., and Nakada, M. 1990. "Effect of Gasoline Quality on Throttle Response of Engines During Warm-Up." SAE Paper No. 900163. SAE International, Warrendale, PA.

3.48 Alquist, H.E., Holman, G.E., and Wimmer, D.B. 1975. "Some Observations of Factors Affecting ORI." SAE Paper No. 750932. SAE International, Warrendale, PA.

3.49 Eng, K.D., Carlson, C.A., Haydn, T.E., and Sung, R.L. 1989. "Engine Test Procedures to Evaluate Octane Requirement Increase and Intake System Cleanliness." SAE Paper No. 892122. SAE International, Warrendale, PA.

3.50 Nishizaki, T., Maeda, Y., Date, K., and Maeda, T. 1979. "The Effect of Fuel Composition and Fuel Additives on Intake System Detergency of Japanese Automobile Engine." SAE Paper No. 790203. SAE International, Warrendale, PA.

3.51 Bond., T.J., Gerry, F.S., and Wagner, W. 1989. "Intake Valve Deposit Control—A Laboratory Program to Optimize Fuel/Additive Performance." SAE Paper No. 892115. SAE International, Warrendale, PA.

3.52 Bitting, W., Gschwendtner, F., Kohlepp, W., Kothe, M., Testvoet, C.J., and Ziwica, K.H. 1987. "Intake Valve Deposits—Fuel Detergency Requirements Revisited." SAE Paper No. 872117. SAE International, Warrendale, PA.

3.53 Shilbolm, C.M., and Schoonveld, G.A. 1990. "Effect on Intake Valve Deposits of Ethanol and Additives Common to the Available Ethanol Supply." SAE Paper No. 902109. SAE International: Warrendale, PA.

3.54 Tupa, R.C., and Koehler, D.E. 1988. "Intake Valve Deposits—Effects of Engines, Fuels and Additives." SAE Paper No. 881645. SAE International, Warrendale, PA.

3.55 Dumont, R., Cunningham, L., Oliver, M., Studzinski, M., Galante, J.M. 2007. "Controlling Induction System Deposits in Flexible Fuel Vehicles Operating on E85." SAE Paper No. 2007-01-4071. SAE International, Warrendale, PA.

3.56 Mikkonen, S., Niemi, A., Niemi, C., and Niskala, J., 1989. "Effect of Engine Oil on Intake Valve Deposits." SAE Paper No. 892111. SAE International, Warrendale, PA.

3.57 Chapman, E., and Cummings, J. 2010. "Effects of Fuel Corrosion Inhibitors on Powertrain Intake Valve Deposits." SAE Paper 2010-01-0908. SAE International, Warrendale, PA.

3.58 Gohn, M.R. 1989. "Evaluating Gasoline Additives for Intake Valve Cleanliness Using the Volkswagen Polo Engine." SAE Paper No. 892118. SAE International, Warrendale, PA.

3.59 Daneshgari, P., Borgman, K., and Job, H. 1989. "The Influence of Temperature Upon Gasoline Deposit Build-Up on the Intake Valves." SAE Paper No. 890215. SAE International, Warrendale, PA.

3.60 Daneshgari, P. 1989. "A Test Procedure for Identifying a Gasoline's Deposit Formation Tendency on Intake Valve." SAE Paper No. 892120. SAE International, Warrendale, PA.

3.61 Dimitroff, E., and Johnston, A. 1966. "Mechanism of Induction System Deposit Formation." SAE Paper No. 660784. SAE International, Warrendale, PA.

3.62 Bailey, B.K., Stavinoha, L.L., and Present, D.L. 1989. "Gasoline Deposit Mechanism on Aluminum at Various Temperatures." SAE Paper No. 892117. SAE International, Warrendale, PA.

3.63 Mackney, D.W. et al. 2002. "Reducing Deposits in a DISI Engine." SAE Paper No. 2002-01-2660. SAE International, Warrendale, PA.

3.64 Parsinejad, F., and Biggs, W. 2011. "Direct Injection Spark Ignition Engine Deposit Analysis: Combustion Chamber and Intake Valve Deposits." SAE Paper No. 2011-01-2110. SAE International, Warrendale, PA.

3.65 Ebert, L.B. (Ed.). 1985. *Chemistry of Combustion Chamber Deposits*. Plenum Press, New York.

3.66 *Proceedings of the CRC Workshop on Combustion Chamber Deposits*, Orlando, FL, November 15–17, 1993. CRC, Atlanta, GA.

3.67 Edwards, J.C., and Choate, P.J. 1993. "Average Molecular Structure of Gasoline Engine Combustion Chamber Deposits Obtained by Solid-State ^{13}C, ^{31}P, and ^{1}H Nuclear Magnetic Resonance Spectroscopy." SAE Paper No. 932811. SAE International, Warrendale, PA.

3.68 Price, R.J., Wilkinson, J.P.T., Jones, D.A.J., and Morley, C. 1995. "A Laboratory Simulation and Mechanism for the Fuel Dependence of SI Combustion Chamber Deposit Formation." SAE Paper No. 952445. SAE International, Warrendale, PA.

3.69 Lauer, J.L., and Friel, P.J. 1960. "Some Properties of Carbonaceous Deposits Accumulated in Internal Combustion Engines." *Combustion and Flame* 4:107.

3.70 Cheng, S.S. 1994. "A Physical Mechanism for Deposit Formation in a Combustion Chamber." SAE Paper No. 941892. SAE International, Warrendale, PA.

3.71 Nakic, D.J., Assanis, D.N., and White, R.A. 1994. "Effect of Elevated Piston Temperature on Combustion Chamber Deposit Growth." SAE Paper No. 940948. SAE International, Warrendale, PA.

3.72 Nagao, M., Kaneko, T., Omata, T., Iwamoto, S., Ohmori, H., and Matsuno, S. 1995. "Mechanism of Combustion Chamber Deposit Interference and Effects of Gasoline Additives on CCD Formation." SAE Paper No. 950741. SAE International, Warrendale, PA.

3.73 Shore, L.B., and Ockert, K.F. 1958. "Combustion Chamber Deposits—A Radio Tracer Study." *SAE Transactions* 66:285–294.

3.74 Newby, W. 1958. "Emphasises the Effect of Boiling Point on Deposit Formation." *SAE Transactions* 66:294.

3.75 Hayes, T.K., White, R.A., and Peters, J.E. 1992. "The In-Situ Measurement of Thermal Diffusivity of Combustion Chamber Deposits in a Spark Ignition Engine." SAE Paper No. 920513. SAE International, Warrendale, PA.

3.76 Hafnan, M., and Nishiwaki, K. 1993. "Determination of Thermal Conductivity and Diffusivity of Engine Combustion Chamber Deposits." JSAE Technical Paper No. 9305931, JSAE, Tokyo.

3.77 Dumont, L.F. 1951. "Possible Mechanisms by Which Combustion Chamber Deposits Accumulate and Influence Knock." *SAE Quarterly Transactions* 5:565–576.

3.78 Nakamura, Y., Yonekawa, Y., and Okamoto, N. 1985. "The Effect of Combustion Chamber Deposits on Octane Requirement Increase and Fuel Economy." In *Chemistry of Combustion Chamber Deposits*, edited by L.B. Ebert, Plenum Press, New York, pp. 199–211.

3.79 Chapman, J.L., Williamson, J., and Preston, W.H. 1989. "Deposits in Internal Combustion Engines." In *Deposition from Combustion Gases*, edited by A.R. Jones, IOP, Bristol, UK, pp. 113–127.

3.80 Cheng, S.S., and Kim, C. 1990. "Effect of Engine Operating Parameters on Engine Combustion Chamber Deposits." SAE Paper No. 902108. SAE International, Warrendale, PA.

3.81 Gebhardt, L.A., Lunt, R.S., and Silbernagel, B.G. 1985. "Electron Spin Resonance Studies of Internal Combustion Engine Deposits." In *Chemistry of Combustion Chamber Deposits*, edited by L.B. Ebert, Plenum Press, New York, pp. 145–175.

3.82 Moore, S.M. 1994. "Combustion Chamber Deposit Interference Effects in Late Model Vehicles." SAE Paper No. 940385. SAE International, Warrendale, PA.

3.83 Kunc, J.F. 1951. "Effect of Lubricating Oil on Octane Requirement Increase." *SAE Quarterly Transactions* 5:582.

3.84 Jackson, M.M., and Pocinki, S.B. 1994. "Effects of Fuel and Additives on Combustion Chamber Deposits." SAE Paper No. 941890. SAE International, Warrendale, PA.

3.85 Kalghatgi, G.T. 1997. "Fuel and Additive Effects on the Rates of Growth of Combustion Chamber Deposits in a Spark-Ignition Engine." SAE Paper No. 972841. SAE International, Warrendale, PA.

3.86 Lepperhoff, G., and Houben, M. 1993. "Mechanisms of Deposit Formation in Internal Combustion Engines and Heat Exchangers." SAE Paper No. 931032. SAE International, Warrendale, PA.

3.87 Anderson, C.L., and Wood, B.S. 1993. "Gasification of Porous Combustion Chamber Deposits in a Spark Ignition Engine." SAE Paper No. 930773. SAE International, Warrendale, PA.

3.88 Hopwood, A.B., Chynoweth, S., and Kalghatgi, G.T. 1998. "A Technique to Measure Thermal Diffusivity and Thickness of Combustion Chamber Deposits In-Situ." SAE Paper No. 982590. SAE International, Warrendale, PA.

3.89 Kalghatgi, G.T. 1996. "Combustion Chamber Deposits and Knock in a Spark-Ignition Engine—Some Additive and Fuel Effects." SAE Paper No. 962009. SAE International, Warrendale, PA.

3.90 Takei, Y., Uehara, T., Hoshi, H., and Okada, M. 1994. "Effects of Gasoline and Gasoline Detergents on Combustion Chamber Deposit Formation." SAE Paper No. 941893. SAE International, Warrendale, PA.

3.91 Duckworth, J.B. 1951. "Effects of Combustion Chamber Deposits on Octane Requirement and Engine Power Output." *SAE Quarterly Transactions* 5:577–583.

3.92 Cornetti, G., Liguori, V., Zanoni, G., and Amendola, L. 1971. "Power Loss Due to Combustion Chamber Deposits." *Journal of Automotive* Engineering 2:8–14.

3.93 Myers, P.S., Uyehara, O.A., and DeYoung, R. 1981. "Fuel Composition and Vaporization Effects on Combustion Chamber Deposits." Report No. DOE/CS/50020-1, U.S. Department of Energy, Washington, DC.

3.94 Kim, C., Cheng, S.S., and Majorski, S.A. 1991. "Engine Combustion Chamber Deposits: Fuel Effects and Mechanisms of Formation." SAE Paper No. 912379. SAE International, Warrendale, PA.

3.95 Choate, P.J., and Edwards, J.C. 1993. "Relationships Between Combustion Chamber Deposits, Fuel Composition and Combustion Chamber Deposit Structure." SAE Paper No. 932812. SAE International, Warrendale, PA.

3.96 Megnin, M.K., and Furman, J.B. 1992. "Gasoline Effects on Octane Requirement Increase and Combustion Chamber Deposits." SAE Paper No. 92258. SAE International, Warrendale, PA.

3.97 Peyla, R.J. 1991. "Motor Gasoline and Deposit Control Additives—A Challenge for the 90s." Paper No. FL-91-118, presented at the 1991 NPRA National Fuels and Lubricants Meeting, Houston, TX.

3.98 Gibbs, L.M. 1993. In *Proceedings of the CRC Workshop on Combustion Chamber Deposits*, Orlando, FL, November 15–17. CRC, Atlanta, GA, p. 7-1.

3.99 Kalghatgi, G.T., Sutkowski, A., Pace, S., Schwahn, H., and Nierhauve, B. 2000. "ASTM Unwashed Gum and the Propensity of a Fuel to Form Combustion Chamber Deposits." SAE Paper No. 2000-01-2026. SAE International, Warrendale, PA.

3.100 Adams, K.M., and Baker, R.E. 1985. "Effects Of Combustion Chamber Deposits Location and Composition." In *Chemistry of Combustion Chamber Deposits*, edited by L.B. Ebert, Plenum Press, New York, pp. 19–37.

3.101 Megnin, M.K., and Choate, P.J. 1994. "Combustion Chamber Deposit Measurement Technique." SAE Paper No. 940346. SAE International, Warrendale, PA.

3.102 Ebert, L.B., Davis, W.H., Mills, D.R. (Part I); Dannerlein, D.I., and Rose, D.L. (Part II); Melchior, M.T. (Part III). 1985. "The Chemistry of Internal Combustion Engine Deposits—Parts I, II, and III." In *Chemistry of Combustion Chamber Deposits*, edited by L.B. Ebert, Plenum Press, New York.

3.103 Kelemann, S.R., and Maxey, C.T. 1993. In *Proceedings of the CRC Workshop on Combustion Chamber Deposits*, Orlando, FL, November 15–17. CRC, Atlanta, GA, p. 7-125.

3.104 Bachman, H.E., and Prestridge, E.B. 1975. "The Use of Combustion Deposit Analysis for Studying Lubricant Induced ORI." SAE Paper No. 750938. SAE International, Warrendale, PA.

3.105 Keller, B.D., Meguerin, G.H., Tracy, C.B., and Smith, J.B. 1976. "ORI of Today's Vehicles." SAE Paper No. 760195. SAE International, Warrendale, PA.

3.106 Harder, R.F., and Anderson, C.L. 1988. "Investigation of Combustion Chamber Deposit Thermal Behavior Utilizing Optical Radiation Measurements in a Fired Engine." *Combustion Science and Technology* 60:423–439.

3.107 Hayes, T.K., Peters, J.E., and White, R.A. 1989. "The Thermal Properties of Engine Deposits and Their Effect on Combustion Chamber Heat Transfer." Paper no. C382/011. In *Proceedings of the 2nd International Conference on New Developments in Powertrain and Chassis Engineering*, Strasbourg, June 1989, Institute of Mechanical Engineers, London.

3.108 Nishiwaki, K., and Hafnan, M. 2000. "The Determination of Thermal Properties of Engine Combustion Chamber Deposits." SAE Paper No. 2000-01-1215. SAE International, Warrendale, PA.

3.109 Kalghatgi, G.T., McDonald, C.R., and Hopwood, A.B. 1995. "An Experimental Study of Combustion Chamber Deposits and Their Effects in a Spark-Ignition Engine." SAE Paper No. 950680. SAE International, Warrendale, PA.

3.110 Cheng, S.S. 1993. "The Effects of Engine Oils on Intake Valve Deposits and Combustion Chamber Deposits." SAE Paper No. 932810. SAE International, Warrendale, PA.

3.111. Benson, J.D. 1975. "Some Factors Which Affect Octane Requirement Increase." SAE Paper No. 750933. SAE International, Warrendale, PA.

3.112 McNab, J.B., Moody, L.E., and Hakala, N.V. 1954. "Effect of Lubricant Composition on Combustion Chamber Deposits." *SAE Transactions* 62:228–242.

3.113 Owrang, F., Mattson, H., Nordlund, A., Olsson, J., and Pedersen, J. 2003. "Characterization of Combustion Chamber Deposits from a Gasoline Direct Injection SI Engine." SAE Paper No. 2003-01-0546. SAE International, Warrendale, PA.

3.114 Pinto da Costa, J.M.C., Sarkisov, L., Seaton, N.A., and Cracknell, R.F. 2009. "Adsorption-Based Structural Characterisation of Combustion

Chamber Deposits." SAE Paper No. 2009-01-0502. SAE International, Warrendale, PA.

3.115 de Goede, S., Rabe, T., Bekker, R., Mtongana, S., and Edwards, J. 2010. "Characterisation of Combustion Chamber Deposits Formed in Direct Injection Spark Ignition (DISI) Engines during an On-Road Vehicle Trial." SAE Paper No. 2010-01-2155. SAE International, Warrendale, PA.

3.116 Arters, D.C., and Macduff, M.J. 2000. "The Effect on Vehicle Performance of Injector Deposits in a Direct Injection Gasoline Engine." SAE Paper No. 2000-01-2021. SAE International, Warrendale, PA.

3.117 Lindgren, R., Skogsberg, M., Sandquist, H., and Denbratt, I. 2003. "The Influence of Injector Deposits on Mixture Formation in a DISC SI Engine." SAE Paper No. 2003-01-1771. SAE International, Warrendale, PA.

3.118 Gething, J.A. 1987. "Performance Robbing Aspects of Intake Valve and Port Deposits." SAE Paper No. 872116. SAE International, Warrendale, PA.

3.119 Amberg, G.H., and Craig, W.S. 1962. "Gasoline Detergents Control Intake System Deposits." SAE Paper No. 554D. SAE International, Warrendale, PA.

3.120 Arters, D.C., Schiferl, E.A., and Szappanos, G. 2002. "Effects of Gasoline Driveability Index, Ethanol and Intake Valve Deposits on Engine Performance in a Dynamometer-Based Cold Start and Warmup Procedure." SAE Paper No. 2002-01-1639. SAE International, Warrendale, PA.

3.121 Costa, J., Sarkisov, L., Seaton, N., and Cracknell, R. 2011. "Adsorption-Based Structural Characterization of Intake Valve Deposits." SAE Paper No. 2011-01-0901. SAE International, Warrendale, PA.

3.122 Spink, C.D., Barraud, P.G., Morris, G.E.L. 1991. "A Critical Road Test Evaluation of Two High-Performance Gasoline Additive Packages in a Fleet of Modern European and Japanese Vehicles." SAE Paper No. 912393 SAE International, Warrendale, PA.

3.123 Spink, C.D., Coleman, A.T., Nelson, E.C., Wakefield, S., and Schoffner, B.A. 2009. "Multi-Vehicle Evaluation of Gasoline Additive Packages: A Fourth Generation Protocol for the Assessment of Intake System Deposit Removal." SAE Paper No. 2009-01-2635. SAE International, Warrendale, PA.

3.124 Price, R.J., Martin, D.P., Dickens, N., and Bohr. 2007. "The Impact of Inlet Valve Deposits on PFI Gasoline SI Engines—Quantified Effects on Fuel Consumption." SAE Paper No. 2007-01-0004. SAE International, Warrendale, PA.

3.125 Yonekawa, Y., Nakamura, Y., and Okamoto, N. 1982." The Study of Combustion Chamber Deposit (Part 4)—Octane Number Requirement and Fuel Economy with Deposit Build Up." *Journal of the Japan Petroleum Institute* 25(3):173–176.

3.126 Yonekawa, Y., Kokubo, K., Nakamura, Y., and Okamoto, N. 1982. "The Study of Combustion Chamber Deposit (part 5): The Role of Combustion Chamber Deposits in Fuel Economy." *Journal of the Japan Petroleum Institute* 25(3):177–182.

3.127 Warren, J. 1954. "Combustion Chamber Deposits and Octane Number Requirement." *SAE Transactions* 62:583–594.

3.128 Harrow, G.A., and Orman, P.L. 1966. "Study of Flame Propagation and Cyclic Dispersion in a Spark Ignition Engine." In *Advances in Automobile Engineering*, edited by G.H. Tidbury, Pergamon Press, Oxford, UK, pp. 3–27.

3.129 Gibson H.J., Hall, C.A., and Huffman, A.E. 1952. "Combustion Chamber Deposition and Power Loss." *SAE Quarterly Transactions* 6(4):595.

3.130 Bradish, J.P., Myers, P.S., and Uyehara, O.A. 1966. "Effects of Deposit Properties on Volumetric Efficiency, Heat Transfer and Preignition in Internal Combustion Engines." SAE Paper No. 660130. SAE International, Warrendale, PA.

3.131 De Gregoria, A.J. 1982. "A Theoretical Study of Engine Deposit and Its Effect on Octane Requirement Using an Engine Simulation." SAE Paper No. 820072. SAE International, Warrendale, PA.

3.132 Nishiwaki, K. 1988. "Unsteady Thermal Behaviour of Engine Combustion Deposits." SAE Paper No. 881225. SAE International, Warrendale, PA.

3.133 Valtadoros, T.H., Wong, V.W., and Heywood, J.B. 199. "Fuel Additive Effects on Deposit Build-Up and Engine Operating Characteristics." *Proceedings of the Symposium on Fuel Composition/Deposit Formation Tendencies*, Atlanta, April 1991. Division of Petroleum Chemistry, ACS, p. 66.

3.134 Bussovansky, S., Heywood, J.B., and Keck, J.C. 1992. "Predicting the Effects of Air and Coolant Temperature, Deposits, Spark Timing and Speed on Knock in Spark Ignition Engines." SAE Paper No. 922324. SAE International, Warrendale, PA.

3.135 Studzinski, W.M., Liiva, P.M., Choate, P.J., Acker, W.P., Litzinger, T., Bower, S., Smooke, M., and Brezinsky, K. 1993. "A Computational and Experimental Study of Combustion Chamber Deposit Effects on NOx Emissions." SAE Paper No. 932815. SAE International, Warrendale, PA.

3.136 Niles, H.T., McConnell, R.J., Roberts, M.A., and Saillant, R. 1975. "Establishment of ORI Characteristics as a Function of Selected Fuels and Engine Families." SAE Paper No. 750451. SAE International, Warrendale, PA.

3.137 Saillant, R.B., Pedrys, F.J., and Kidder, H.E. 1976. "More Data on ORI Variables." SAE Paper No. 760196. SAE International, Warrendale, PA.

3.138 Schreyer, P., Starke, K., Thomas, J., and Crema, S. 1993. "Effect of Multifunctional Fuel Additives on Octane Number Requirement of Internal Combustion Engines." SAE Paper No. 932813. SAE International, Warrendale, PA.

3.139 Graiff, L.B. 1979. "Some New Aspects of Deposit Effects on Engine Octane Requirement Increase and Fuel Economy." SAE Paper No. 790938. SAE International, Warrendale, PA.

3.140 Tsutsumi, Y., Nomura, K., and Nakamura, N. 1990. "Effect of Mirror-Finished Combustion Chamber on Heat Loss." SAE Paper No. 902141. SAE International, Warrendale, PA.

3.141 Heywood, J.B. 1988. "Engine Design and Operating Parameters" (chap. 2). In *Internal Combustion Engine Fundamentals*, J.B. Heywood McGraw-Hill, New York, pp. 42–61.

3.142 Hoard, J., and Moilanen, P. 1997. "Exhaust Valve Seat Leakage." SAE Paper No. 971638. SAE International, Warrendale, PA.

3.143 Kalghatgi, G.T., and Price, R.J. 2000. "Combustion Chamber Deposit Flaking." SAE Paper No. 2000-01-2858. SAE International, Warrendale, PA.

3.144 Kalghatgi, G.T. 2002. "Combustion Chamber Deposit Flaking—Studies Using a Road Test Procedure." SAE Paper No. 2002-01-2833. SAE International, Warrendale, PA.

3.145 Kalghatgi, G.T. 2003. "Combustion Chamber Deposit Flaking and Startability Problems in Three Different Engines." SAE Paper No. 2003-01-3187. SAE International, Warrendale, PA.

3.146 Bitting, W.H., Firmstone, G.P., and Keller, C.T. 1994. "Effects of Combustion Chamber Deposits on Tailpipe Emissions." SAE Paper No. 940345. SAE International, Warrendale, PA.

3.147 Gagliardi, J.C. 1967. "The Effects of Fuel Anti-Knock Compounds and Deposits on Exhaust Emissions." SAE Paper No. 670128. SAE International, Warrendale, PA.

3.148 Huls, T.A., and Nickol, H.A. 1967. "Influence of Engine Variables on Exhaust Oxides of Nitrogen Concentrations from a Multicylinder Engine." SAE Paper No. 670482. SAE International, Warrendale, PA.

3.149 Harpster, M.O., Matas, S.E., Fry, J.H., and Litzinger, T.A. 1995. "An Experimental Study of Fuel Composition and Combustion Chamber Deposit Effects on Emissions from a Spark Ignition Engine." SAE Paper No. 950740. SAE International, Warrendale, PA.

3.150 Bitting, W.H., Firmstone, G.P., and Keller, C.T. 1995. "A Fleet Test of Two Additive Technologies Comparing Their Effects on Tailpipe Emissions." SAE Paper No. 950745. SAE International, Warrendale, PA.

3.151 Watanabe, M. et al. 2004. "Effects of CCD on Emissions from DISI Engine Using Different Fuel Distillation Properties." SAE Paper No. 2004-01-1954. SAE International, Warrendale, PA.

3.152 Wentworth, J.T. 1971. "Effect of Combustion Chamber Surface Temperature on Exhaust Hydrocarbon Concentration." SAE Paper No. 710587. SAE International, Warrendale, PA.

3.153 Myers, J.P., and Alkidas, A.C. 1978. "Effects of Combustion-Chamber Surface Temperature on the Exhaust Emissions of a Single-Cylinder Spark-Ignition Engine." SAE Paper No. 780642. SAE International, Warrendale, PA.

3.154 Russ, S.G., Kaiser, E.W., Siegl, W.O., Podsiadlik, D.H., and Barrett, K.M. 1995. "Compression Ratio and Coolant Temperature Effects on HC Emissions from a Spark-Ignition Engine." SAE Paper No. 950163. SAE International, Warrendale, PA.

3.155 Adamczyck, A.A., and Kach, R.A. 1986. "The Effect of Engine Deposit Layers on Hydrocarbon Emissions from Closed Vessel Combustion." *Combustion Science and Technology* 47:193–212.

3.156 Haidar, H.A., and Heywood, J.B. 1997. "Combustion Chamber Deposit Effects on Hydrocarbon Emissions from a Spark-Ignition Engine." SAE Paper No. 972887. SAE International, Warrendale, PA.

3.157 Leikkanen, H.E., and Beckman, E.W. 1971. "The Effect of Leaded and Unleaded Gasolines on Exhaust Emissions as Influenced by Combustion Chamber Deposits." SAE Paper No. 710843. SAE International, Warrendale, PA.

3.158 Pahnke, A.J., and Conte, J.F. 1969. "Effect of Combustion Chamber Deposits and Driving Conditions on Vehicle Exhaust Emissions." SAE Paper No. 690017. SAE International, Warrendale, PA.

3.159 Bower, S.L., Litzinger, T.A., and Frottier, V. 1993. "The Effects of Fuel Composition and Engine Deposits on Emissions from a Spark Ignition Engine." SAE Paper No. 932707. SAE Paper No. 690017. SAE International, Warrendale, PA.

3.160 Russ, M.J. 1993. In *Proceedings of the CRC Workshop on Combustion Chamber Deposits*, Orlando, FL, November 15–17. CRC, Atlanta, GA, pp. 4–31.

3.161 Greenshields, R.J. 1953. "Spark Plug Fouling Tendencies." *SAE Transactions* 61:3.

3.162 Kimbara, Y., Noguchi, Y., and Ishiguro, T. 1980. "Study on Spark Plug Carbon Fouling." SAE Paper No. 800832. SAE International, Warrendale, PA.

3.163 Quader, A. 1992. "Spark Plug Fouling: A Quick Engine Test." SAE Paper No. 920006. SAE International, Warrendale, PA.

3.164 Collings, N., Dinsdale, S., and Hands, T. 1991. "Plug Fouling Investigations on a Running Engine—an Application of a Novel Multi-Purpose Diagnostic System Based on the Spark Plug." SAE Paper No. 912318. SAE International, Warrendale, PA.

3.165 Kalghatgi, G.T. 1988. "Improvements in the Ignition Ability of Glow Discharges Brought About by Deposits of Potassium Sulphate on the Electrodes." Paper No. C50/80, presented at the *International Conference on Combustion in Engines—Technology and Applications*, Institute of Mechanical Engineers, London.

3.166 Kalghatgi, G.T. 1989. "Flame Initiation and Development from Glow Discharges—Effect of Electrode Deposits." *Combustion and Flame* 77:321.

3.167 Kalghatgi, G.T. 1987. "Improvement in Early Flame Development in a Spark Ignition Engine Brought About by a Spark-Aider Fuel Additive." *Combustion Science and Technology* 52:427.

3.168 Spink, C.D., McDonald, C.R., Morris, G.E.L., and Stephenson, T. 2004. "A Critical Road Test Evaluation of a High-Performance Gasoline Additive Package in a Fleet of Modern European and Asian Vehicles." SAE Paper No. 2004-01-2027. SAE International, Warrendale, PA.

3.169 Kalghatgi, G.T. 1988. "Effect of Spark-Aider Fuel Additives on the Misfire Characteristics of a Spark Ignition Engine." *Combustion Science and Technology* 62:1–19.

3.170 Tupa, R.C., Dorer, C.J. 1986. "Gasoline and Diesel Fuel Additives for Performance/ Distribution Quality—II." SAE Paper No. 861179. SAE International, Warrendale, PA.

3.171 Kipp, K.L., Ingamells, J.C., Richardson, W.L., and Davis, C.E. 1970. "Ability of Gasoline Additives to Clean Engines and Reduce Exhaust Emissions." SAE Paper No. 700456. SAE International, Warrendale, PA.

3.172 Udelhofen, J.M., and Zahalka, T.L. 1988. "Gasoline Additive Requirements for Today's Smaller Engines." SAE Paper No. 881644. SAE International, Warrendale, PA.

3.173 Aradi, A.A., Evans, J., Miller, K., and Hotchkiss, A. 2003. "Direct Injection Gasoline (DIG) Injector Deposit Control with Additives." SAE Paper No. 2003-01-2024. SAE International, Warrendale, PA.

3.174 Schwahn, H., Lutz, U., and Kramer, U. 2010. "Deposit Formation of Flex Fuel Engines Operated on Ethanol and Gasoline Blends." SAE Paper No. 2010-01-1464. SAE International, Warrendale, PA.

3.175 Bert, J.A. et al. 1983. "A Gasoline Additive Concentrate Removes Combustion Chamber Deposits and Reduces Vehicle Octane Requirements." SAE Paper No. 831709. SAE International, Warrendale, PA.

3.176 Gibbs, L.M., Lewis, R.A., Voss, D.A., and Jones, D.R. 1979. "Chevron Experience with a New Deposit Control Gasoline." Paper No. FL-79-109, presented at the *National Fuels and Lubricants Meeting of the National Petroleum Refiners' Association*, Houston, November 1979.

3.177 Polss, P. 1973. "What Additives Do for Gasoline." *Hydrocarbon Processing* 52(2):61.

3.178 Asseff, P.A. 1966. "Multifunctional Gasoline Additives Reduce Engine Deposits." SAE Paper No. 660543. SAE International, Warrendale, PA.

3.179 Sheahan T.J., Dorer, C.J., and Miller, C.O. 1969. "Detergent-Dispersant Fuel Performance and Handling." SAE Paper No. 690516. SAE International, Warrendale, PA.

3.180 Swain, M.R., Swain, M.N., Blanco, J.G., and Adt, R.R. 1987. "The Effect of Alternative Gasolines on Knock and Intake Valve Sticking." SAE Paper No. 872040. SAE International, Warrendale, PA.

3.181 Mikkonen, S., Karlsson, R., and Jouni, K. 1988. "Intake Valve Sticking in Some Carburetor Engines." SAE Paper No. 881643. SAE International, Warrendale, PA.

3.182 Ho Teng. 2010. "Physicochemical Characteristics of Soot Deposits in EGR Coolers." SAE Paper No. 2010-01-0730. SAE International, Warrendale, PA.

3.183 Abraham, M., et al. 2010. "Review of Soot Deposition and Removal Mechanisms in EGR Coolers." SAE Paper No. 2010-01-1211. SAE International, Warrendale, PA.

3.184 Strots, V.O., et al. 2009. "Deposit Formation in Urea-SCR Systems." SAE Paper No. 2009-01-2780. SAE International, Warrendale, PA.

Chapter 4
Fuel Effects on Autoignition in Premixed Systems—Knock in Spark Ignition Engines and Combustion in Homogeneous Charge Compression Ignition Engines

In both spark ignition (SI) engines and homogeneous charge compression ignition (HCCI) engines, fuel and air are fully premixed and the mixture is compressed. However, the mixture can never be truly homogeneous because there will always be temperature and mixture strength gradients caused by turbulence. Energy release occurs in SI engines as the turbulent flame front initiated by the spark advances through the mixture. As pressure increases in the cylinder, the temperature of the unburned end-gas ahead of the advancing flame also increases and, under some circumstances, could cause the end-gas to autoignite before the flame consumes it.

The autoignition is initiated around one or more ignition centers or "hot spots." The autoignition could cause a very rapid release of much of the chemical energy contained in the end-gas, which could lead to a sharp pressure rise and pressure waves inside the cylinder. It is like hitting the inside of the cylinder with a hammer. The engine vibrates with dominant frequencies of 5–10 kHz depending on the engine geometry and generates the sharp pinging sound characteristic of knock. The intensity of knock is commonly characterized by the maximum amplitude of the pressure fluctuations if the pressure inside the cylinder is measured. Knock can cause engine damage, and the strategies used to manage knock reduce the efficiency and the maximum power of SI engines. In HCCI engines, the entire charge autoignites, whereas in knock in SI engines, only a part of the charge in the end-gas autoignites. Inevitably, the chemical energy that can be released per cycle for a given volume of charge in an HCCI engine will be much lower than in SI engines in order to limit damagingly high-pressure rise rates.

Autoignition depends on the evolution with time of the pressure and temperature of the fuel/air mixture in the end-gas, and on the autoignition chemistry of this mixture, which depends critically on the fuel composition and on the mixture strength. This chapter discusses the fuel effects on autoignition and the consequences for knock in SI engines and combustion in HCCI engines. Autoignition properties of fuels as measured by RON (research octane number), MON (motor octane number), and cetane number (see Chapter 2) are very important in fuel manufacturing, and their relevance for practical fuels is considered.

4.1 Autoignition

Westbrook reviewed the chemical kinetics of autoignition in practical combustion systems [4.1]. Autoignition occurs when chain reactions cause an exponential increase in temperature, leading to the eventual consumption of the fuel to release its chemical energy as sensible energy. It starts with initiation reactions that generate radicals from stable species. Propagation reactions maintain the number of radicals, and chain-branching reactions increase the number of radicals; identifying the chain-branching steps under the conditions being studied is essential for understanding the chemical kinetics [4.1]. Finally, there are termination reactions, which reduce the number of radicals.

The overall reaction rate reflects the combined effect of all these reactions, which might number in the thousands. The pressure and temperature vary over a wide range in internal combustion (IC) engines, and different reactions become important at different pressures and temperatures. Westbrook [4.1] argued that autoignition in engines is dominated by one reaction, that is, the decomposition of H_2O_2 at a temperature of around 900–1000 K, which produces OH radicals. This temperature range, when the main heat release occurs, could be lower under high-pressure conditions. For instance, in the HCCI experiments described in [4.2], for the same mixture of iso-octane and n-heptane, at 2 bar (0.2 MPa) absolute intake pressure and 40 °C intake temperature, final heat release occurred at a temperature of around 760 K; whereas with the intake pressure at 1 bar (0.1 MPa) absolute and intake temperature of 120 °C, the final heat release occurred at around 960 K (see Fig. 9 in [4.2]). The chemical reactions at lower temperatures and pressures determine when this final stage is reached during the engine cycle.

4.1.1 Chemical Kinetics Modeling

Comprehensive chemical models of autoignition aim to use all relevant chemical reactions. Westbrook, Dryer, Warnatz, and colleagues have built reaction mechanisms consisting of thousands of reactions for combustion chemistry, starting with simple fuels and moving on to larger molecules [e.g., 4.3–4.5]. However, "comprehensive" schemes cannot be perfectly accurate because of the uncertainties in the reaction rate constants and their temperature and pressure dependences [4.5, 4.6]. Thus, comprehensive models include all available, though imperfect, chemical knowledge. They have to be calibrated using experimental data and can be used to interpolate and extrapolate chemical behavior. Many practical applications use reduced or simplified chemical models that attempt to select the reactions of critical importance for the particular application [4.7]. A brief discussion of such models can also be found in [4.6]. Such models inevitably have a limited representation of real chemistry. However, for a given fuel and a given application, these models can be tuned by comparison with experiments. Then they can be used to predict autoignition phenomenon such as knock at other experimental conditions.

Chemical reaction mechanisms applicable to autoignition in IC engines have only been developed for a very few pure compounds—mostly paraffins of low carbon numbers. Simmie [4.5] has reviewed the literature on such models for different hydrocarbon fuels. Chemical kinetic schemes have been developed for mixtures of iso-octane and n-heptane, known as primary reference fuels (PRFs), but practical fuels contain many chemical compounds (Chapter 2) whose chemistry is very different from that of PRFs. Development of models that represent all these components is not possible because the model would be too large for current computational resources, and because the fundamental data needed for development of such a model (e.g., chemical kinetic rate constants, reaction paths, and thermodynamic parameters) are not available.

Thus, simplified surrogate fuels are needed for representing practical fuels [4.8, 4.9]. A *surrogate fuel* is defined as a fuel composed of a small number of pure compounds and whose behavior matches the target practical fuel in terms of combustion and emissions characteristics. The chemical kinetic mechanisms for such surrogates are being developed [e.g., 4.10–4.13]. However, predicting autoignition, even in engines where the fuel and air are fully mixed, requires such kinetic models to be coupled to models that predict turbulent flow and mixture preparation in the engine and, for SI engines, turbulent flame development. The full kinetic models for surrogates are too large to be used in such computational engine models; a model for a surrogate with five components in [4.10] had 1328 species and 5835 reactions. Andrae et al. [4.12] described a semidetailed mechanism for a mixture of PRF and toluene with 137 species and 633 reactions, but even this is too large to be coupled to an engine model. Experiments have to play a central role in the study of autoignition.

4.1.2 Ignition Delay and the Livengood-Wu Integral

Experiments conducted in a rapid compression machine or a shock tube form the bedrock for autoignition studies. In these devices the fuel/oxidant mixture is compressed rapidly and held at a nominally constant pressure and temperature. The time between the end of compression and autoignition, as marked by heat release, is known as the induction time or ignition delay τ_i. The lower the value of τ_i, the more reactive is the mixture. Quite often ignition takes place in two stages, with a smaller, earlier heat release associated with low-temperature reactions [4.1]. For a given fuel, the ignition delay changes with temperature and pressure and depends on the equivalence ratio. Chemical kinetic models of autoignition are usually calibrated using experimental values of ignition delay. In IC engines the pressure and temperature of the fuel/air mixture both increase from the start of compression, and at each point during the cycle, the mixture will have a different value of τ_i. It is then hypothesized that autoignition occurs when the integral I of the reciprocal of τ_i, with time t, reaches unity (Eq. 4.1):

$$I = \int_0^{te} \frac{dt}{\tau_i(P,T)} = 1 \tag{4.1}$$

This hypothesis was first suggested by Livengood and Wu [4.14] and is now commonly used in knock studies as well as HCCI studies [e.g., 4.15, 4.16]. The integral I represents the fraction of the reactions needed for autoignition to be achieved completed by time te. It is very important to account for the effects of changing pressure P. An increase in pressure causes a decrease in τ_i—the mixture becomes more reactive. In general, τ_i can be expressed over a given temperature range as (Eq. 4.2):

$$\frac{\tau_i}{\tau_{i0}} = f\left(\frac{T}{T_0}\right)\left(\frac{P}{P_0}\right)^{-n} \tag{4.2}$$

Here, τ_{i0} is the ignition delay measured at a pressure and temperature of P_0 and T_0, respectively (see Figs. 4.1 and 4.2). Figure 4.1 shows τ_i measured for iso-octane for different pressures and temperatures; the data are replotted from Fig. 11 of [4.17]. The data can be collapsed to a single function of temperature if they are scaled to a single pressure using Eq. 4.2, as illustrated in Fig. 4.2, which shows τ_i scaled to 40 bar (4 MPa) by multiplying each value by a scaling factor $(40/P)^{-1.5}$—in this case, n was chosen by trial and error to be 1.5.

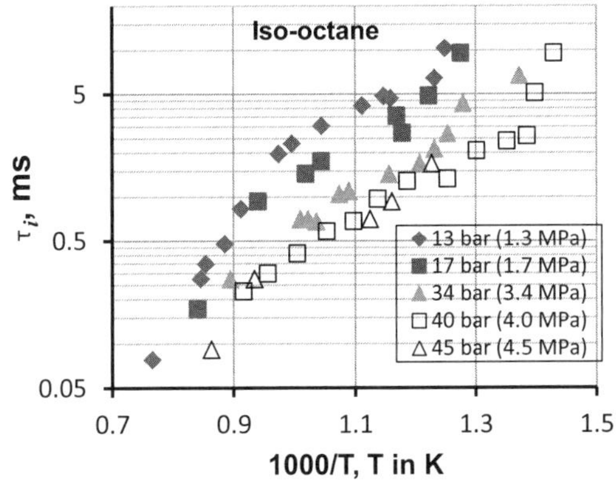

Fig. 4.1. Ignition delay for iso-octane at stoichiometric conditions at different pressures and temperatures from the shock tube studies of Fieweger et al. (Fig. 11 of [4.17]).

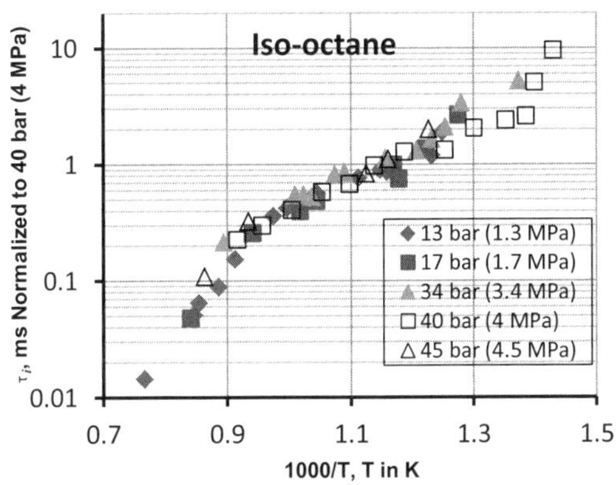

Fig. 4.2. Ignition delay in Fig. 4.1 scaled to 40 bar (4 MPa) by multiplying each value of τ_i by $(40/P)^{-1.5}$, where P is the actual pressure in bar at which τ_i was measured.

The value of the pressure exponent n can be different in different temperature ranges and also depends on the equivalence ratio. However, only low values of ignition delay (<1 ms) in the higher temperature and pressure range, such as considered in Fig. 4.2, contribute significantly to the Livengood-Wu integral. The value of n has been experimentally determined for different fuels in this critical pressure and temperature range by several groups. Gauthier et al. [4.18] measure n to be 1.64 for n-heptane. Hence, we should expect the pressure exponent n for PRF to be between 1.5 and 1.64. However, for nonparaffinic fuels, the value of n has been measured to be around 1.0 or less [4.18–4.20]. Table 4.1 lists the measured values of n for different fuels, along with the

Table 4.1 Measured Values of the Pressure Exponent *n* for Different Fuels. Stoichiometric Mixture Strength, Φ = 1.

	RON	MON	iso-octane vol%	n-heptane vol%	toluene vol%	ethanol vol%	DIB** vol%	*n*	Ref.
Toluene reference fuel (TRF) 84***	84	73		35	65			1.06	4.19
Fuel B	95.1	89.5	62	18		20		0.76	4.20
Fuel C	94.6	85	25	20	45		10	0.65	4.20
Gasoline*	92	82						1.05	4.18
Surrogate A	88	85	63	17	20			0.83	4.18
Surrogate B	87	85	69	17	14			0.96	4.18
n-heptane	0	0		100				1.64	4.18
iso-octane	100	100	100					1.5	Fig. 4.2

*(RON + MON)/2 for gasoline was 87, assumed sensitivity of 10. RON and MON for Surrogate A and Surrogate B were estimated from blending rules.
**DIB is di-isobutylene, 2-4-4 trimethyl-1-pentene.
***Fuel A in [4.20].

composition of these fuels. The composition of gasoline considered in Table 4.1 can be found in [4.18].

Thus, for a given temperature, if the pressure is raised, the value of τ_i decreases less for non-PRFs (i.e., they become relatively more resistant to autoignition compared to PRFs). In an engine both pressure and temperature change in the lead-up to autoignition, and hence the mixture does not follow a single variation with temperature at a fixed pressure. For instance, in an SI engine running on iso-octane at stoichiometric conditions, the end-gas temperature at 20 bar (2 MPa) could be 700 K. As the flame front develops, the pressure in the combustion chamber rises, say, to 50 bar (5 MPa), the temperature would be around 900 K. The mixture would have moved from point A to point B in Fig. 4.3, where the data for iso-octane in Fig. 4.1 have been scaled to 20 bar (2 MPa) and 50 bar (5 MPa), respectively, again using a value of 1.5 for n.

Ignition delay also depends on the mixture strength in addition to pressure and temperature as illustrated in Fig. 4.4 where τ_i for TRF 84 (Table 4.1) at 40 bar (4 MPa) is shown for equivalence ratio, Φ, of 0.3 and 1. The data are taken from Table 1 of [4.19] and have been scaled to 40 bar (4 MPa) using $n = 0.88$ for $\Phi = 0.3$ and $n = 1.06$ for $\Phi = 1$, the values derived in [4.19]. For the leaner mixture, ignition delay is significantly higher.

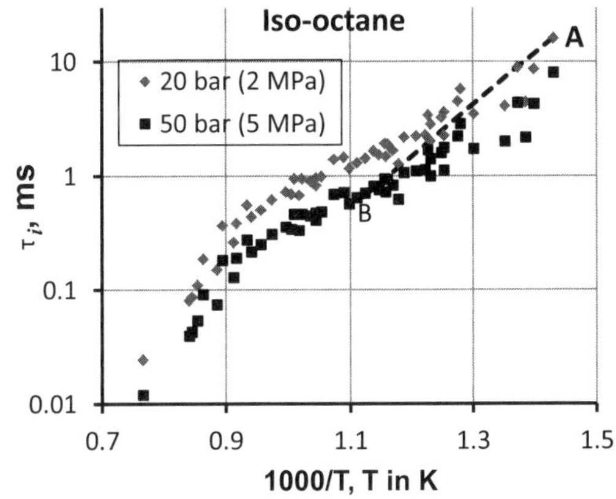

Fig. 4.3. Ignition delays for iso-octane at stoichiometric condition scaled to 50 bar (5 MPa) and 20 bar (2 MPa). As the mixture is compressed from 20 bar (2 MPa) to 50 bar (5 MPa), it moves from A to B.

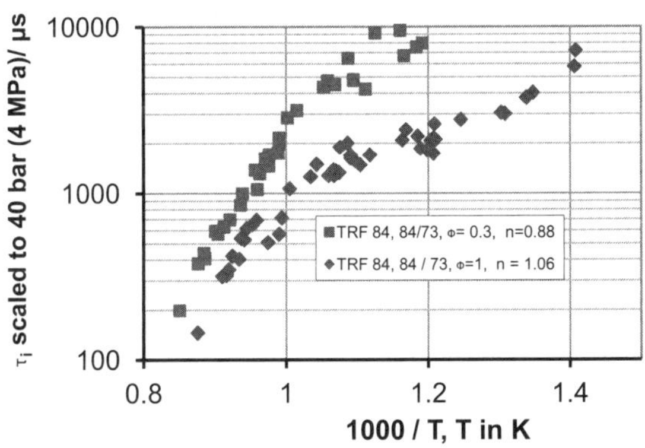

Fig. 4.4. Ignition delays for TRF 84 (Table 4.1) at two different equivalence ratios (Φ). The data from Table 1 of [4.19] have been scaled to 40 bar (4 MPa) using $n = -0.88$ for $\Phi = 0.3$ and $n = -1.06$ for $\Phi = 1$. The ignition delay is very much higher for the leaner mixture.

Modern engines routinely use EGR (exhaust gas recirculation) to control combustion and emissions. Some of the components of EGR gases such as NO can also influence autoignition chemistry [e.g., 4.21–4.24]. The variations with time of pressure, temperature, mixture strength, and mixture composition, as well as the autoignition chemistry of the fuel, need to be known to predict autoignition in the engine. In SI and HCCI engines, the global mixture strength does not change during the cycle, with the equivalence ratio $\Phi = 1$ in modern SI engines and $\Phi < 1$ in HCCI engines.

The functional variation of τ_i with temperature T at a fixed pressure P can be represented by an Arrhenius type expression such as Eq. 4.3, in particular temperature ranges,

$$\tau_i = f(T) = A\exp\left(\frac{-E}{RT}\right) \tag{4.3}$$

Here, E is the activation energy, R is the universal gas constant, and A is a pre-exponential constant. A and E are different for different fuels over a given pressure and temperature range; their values also depend on the equivalence ratio for a given fuel. The ignition delay time at high and low temperatures could well be expressed as different Arrhenius expressions such as Eq. 4.3, but in the temperature range between about 700 and 900 K, τ_i might change very little with temperature for some fuels. Indeed, for some fuels, particularly paraffins, τ_i can actually increase (i.e., reaction rates decrease with increasing temperature; e.g., [4.25]) in this temperature range. This phenomenon is described as the negative-temperature coefficient (NTC) behavior. The heat release often observed before the main heat release in autoignition experiments (two-stage autoignition) is associated with the NTC region, and it is called the first-stage heat release and low-temperature heat release (LTHR) in the HCCI literature. It is common to describe the chemistry in the NTC region as "low-temperature" chemistry. Temperature measurements in the unburned mixture using CARS (coherent anti-Stokes Raman spectroscopy) before the occurrence of knock in cycles that knocked showed a significant increase in temperature above that expected from polytropic compression [4.26] for several different fuels. This increase in temperature, presumably caused by the low-temperature chemistry, was not observed for propane, which is not expected to show any NTC behavior [4.26]. During autoignition in SI and HCCI engines, the fuel/air mixture traverses the NTC region to varying degrees depending on engine design and operating conditions.

The pressure exponent n could also be different in different pressure and temperature ranges. For instance, Gauthier et al. [4.18] measured n to be 1.64 for n-heptane, but it was 0.55 in another study by the same group at a higher temperature and lower pressure range [4.18]. They suggested that this reflects the difference in the NTC chemistry in the two different temperature ranges. The evidence from several different experiments suggests that the NTC region occurs at higher temperatures for higher pressures [4.18]. Yates et al. [4.27] have suggested approximating the dependence of τ_i on T and P with three separate

Arrhenius type expressions combined with different values of n at different temperature ranges. However, in typical engine tests, ignition delays greater than about 10 ms (e.g., beyond the range considered in Fig. 4.2) contribute little to the Livengood-Wu integral. The values of n listed in Table 4.1 cover this range, and the difference in the value of n between different fuels is relevant to practical engine operation.

We now illustrate the implication of different dependences on temperature and pressure of the ignition delay on autoignition in different engine operating conditions using the Livengood-Wu integral (Eq. 4.1). We consider PRF 84 (84% vol iso-octane + 16% vol n-heptane) and TRF 84 from Table 4.1, both of which have been studied in several HCCI engine experiments [4.2, 4.28, 4.29]. Both have the same RON of 84, but TRF 84 has a lower MON of 73. We consider hypothetical HCCI experiments at two operating conditions with the engine running at 1200 rpm at stoichiometric conditions ($\Phi = 1$): Condition 1 with intake pressure P_{in} and intake temperature T_{in} of 2 bar (0.2 MPa) absolute and 80 °C, respectively [4.29], and Condition 2 with P_{in} of 1 bar (0.1 MPa) absolute and T_{in} of 250 °C [4.28]. The pressure (measured) and temperature (estimated) variations with crank angle at these conditions are shown in Fig. 4.5.

The variation of ignition delay with temperature at a fixed pressure of 40 bar (4 MPa) for the two fuels is shown in Fig. 4.6.

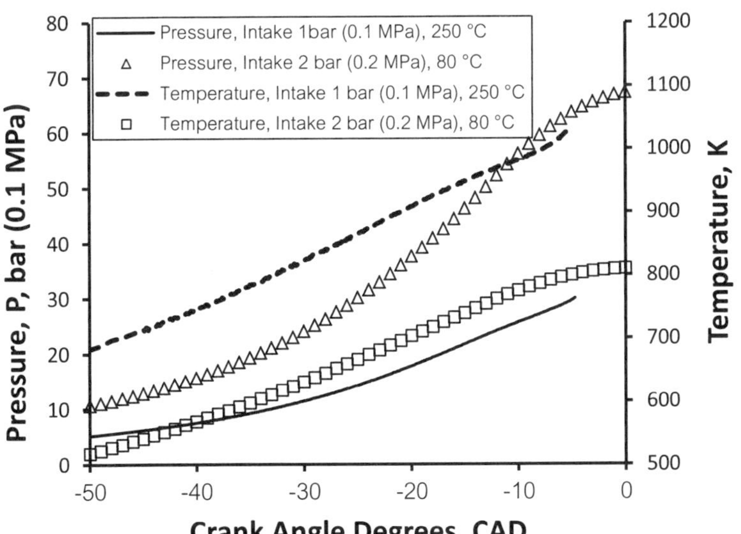

Fig. 4.5. Pressure and temperature variation with crank angle at two operating conditions of an engine running at 1200 rpm [4.28, 4.29]. Condition 1: P_{in} = 2 bar (0.2 MPa) absolute, T_{in} = 80 °C; Condition 2: P_{in} = 1 bar (0.1 MPa) absolute, T_{in} = 250 °C.

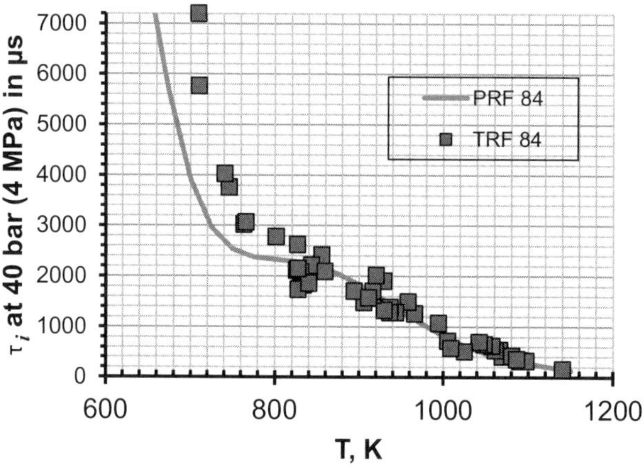

Fig. 4.6. Ignition delay at 40 bar (4 MPa) for PRF 84 and TRF 84 at stoichiometric conditions. The data for PRF 84 from interpolation of results are in Fig. 17 of [4.17]. The data for TRF 84 are from Table 1 of [4.19] scaled to 40 bar (4 MPa) using $n = 1.06$.

The ignition delay for PRF 84 was interpolated from the results in Fig. 17 of [4.17], as in [4.16]; those for TRF 84 are shown in Fig. 4.4. Note that in Fig. 4.6, the scale for ignition delay is linear rather than logarithmic, and it has been plotted against T rather than against $1000/T$. The ignition delays shown in Fig. 4.6 are for stoichiometric conditions, and the discussion that follows is for illustration only, because HCCI experiments with these fuels could only be done at lean conditions with equivalence ratios of less than 0.35 to avoid excessive pressure rise rates. Using Fig. 4.6 and an appropriate value for n (1.6 for PRF 84 and 1.06 for TRF 84), τ_i can be calculated at each crank angle at the two operating conditions since the pressure and temperature are known, and the Livengood-Wu integral I can be estimated. In Fig. 4.7, I has been plotted against crank angle for the two operating conditions for the two fuels.

At Condition 1 with the higher intake pressure (2 bar (0.2 MPa) absolute), I reaches unity for TRF 84 later in the cycle compared to PRF 84 (i.e., TRF 84 is more resistant to autoignition than PRF 84), even though it has a lower MON and the same RON. Hence, at this pressure and temperature condition, TRF 84 will match the PRF of an octane number much higher than 84. Indeed, at such high-pressure conditions, TRF 84 was significantly more resistant to autoignition than PRF 84 in HCCI experiments [4.2]. However, at Condition 2, with a higher intake temperature and lower intake pressure compared to Condition 1, TRF 84 ignites earlier, and thus is less resistant to autoignition than PRF 84,

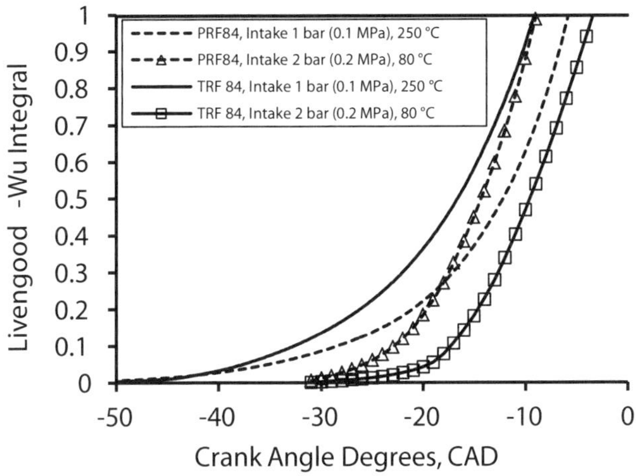

Fig. 4.7. Livengood-Wu integral for PRF 84 and TRF 84 for Conditions 1 and 2.

and will match the PRF of an octane number lower than 84 for autoignition behavior. At a high-temperature and low-pressure case in an HCCI experiment (OP4 in [4.2]), TRF 84 did indeed ignite earlier than PRF 84. If we were to assume the same values for the pressure exponent n for the two fuels at the high-temperature and low-pressure condition, the Livengood-Wu integral for TRF 84 would not reach unity before PRF 84, contrary to experimental observation [4.2]. Also, clearly in the MON test, which has a higher intake temperature than the RON test, TRF 84 with a MON of 73 is less resistant to knock than PRF 84. Thus, the same non-PRF such as TRF 84 can match different PRFs in terms of autoignition behavior, depending on the pressure and temperature history of the mixture because of different dependences of the ignition delay on pressure and temperature.

4.1.3 Autoignition Quality of Practical Fuels

In principle, if τ_i is established accurately for a fuel, either experimentally or through a suitable kinetic model as a function of temperature, pressure, and mixture strength, the Livengood-Wu integral could be used to predict autoignition if the pressure and temperature in the unburned gas mixture in the engine could be accurately predicted. Predicting the pressure and temperature in the unburned mixture before autoignition is easier in an HCCI engine; an accurate flame development model would be needed in an SI engine to predict pressure and temperature in the end-gas once combustion has started. Of course, the kinetic model itself could predict autoignition without bringing in the empiricism associated with the Livengood-Wu integral.

However, current computing capacity is not generally sufficient to handle computational models that integrate turbulent flow, flame development in the SI engine, and kinetic models to predict autoignition. Ignition delay data are difficult to generate, and there are only a few centers in the world that are capable of doing so. Therefore, data over a reasonable pressure and temperature range only exist for a few fuels. Nevertheless, these approaches could be used to study the autoignition of a well-characterized fuel in engines but are not now useful in characterizing autoignition quality of practical fuels in general and in the manufacture and blending of fuels to meet particular autoignition quality targets. For instance, fuels are manufactured and marketed to meet standards based on RON, MON, or cetane number (see Chapter 2). Conceivably, such standards could be replaced by computer models that specify some agreed-upon autoignition criterion that the fuel would meet under conditions such as specified pressure and temperature variations with time. The unknown fuel in question has to have its chemical composition established and a suitable kinetic model devised to be used in the computer model to see if it meets the required quality. Such information has to be developed for all the blending components used to make a fuel, and blending rules have to be established. Such an approach is likely to remain too difficult to implement computationally.

Hence, fuel autoignition quality has to be specified by empirical measures such as the RON and the MON. Of course, RON and MON are far from perfect as quantifiers of fuel autoignition quality, as will be discussed next. However, RON and MON have had more than 80 years to establish themselves across the world and are central to the manufacture of fuels. Autoignition in engines using practical fuels will need to be described and understood in terms of the RON and MON of the fuel in the foreseeable future regardless of any advances in the fundamental understanding of autoignition and the accuracy and the applicability of the models.

4.2 Knock in SI Engines and Fuel Antiknock Quality

On average, knock onset as marked by the start of the pressure fluctuations follows the peak in the heat release rate [4.26] unless there are multiple autoignition events in the same cycle. Not every engine cycle knocks even when knock is observed in engines. Combustion in SI engines is marked by significant cycle-to-cycle variations [e.g., 4.30–4.32], and cycles with faster flame development will have higher pressures and are more likely to knock [4.33]. In the lead-up to knock, knocking cycles have higher burning rates and higher pressures at a given crank angle compared with nonknocking cycles. However, the mixture temperature is higher for knocking cycles only later in the cycle, nearer the final heat release [4.33].

4.2.1 Knock Intensity and Knock-Limited Spark Advance

Knock is the name given to the sharp metallic noise that results from autoignition in the end-gas [4.30]. Knock can also be characterized using more objective measures such as in-cylinder pressure. The fluctuating pressure signal associated with knock is usually extracted by filtering the pressure signal 5–10 kHz using electronic filters. The maximum peak-to-peak amplitude of this signal averaged over many cycles is often used as a measure of knock intensity, though many other pressure-based methods have also been considered [4.34, 4.35]. Knock sensors in vehicles usually measure engine vibrations using an accelerometer, though other methods such as those based on measuring fluctuations in ion current in the spark gap have also been used (which depends on the pressure; e.g., [4.36]). Whatever the method used to measure knock, some threshold level has to be chosen to define if an engine cycle is experiencing knock. In this context autoignition does not always lead to knock. Autoignition can be benign—that is, even when it is optically observed, the resultant maximum heat release rate may be too low to cause pressure fluctuations at the threshold level [4.37, 4.38] or lead to the characteristic noise.

The spark usually fires before top dead center (TDC), the top of the compression stroke, in an SI engine. The pressure in the cylinder increases (e.g., the peak pressure is higher; [4.39]) as the ignition timing is advanced because of better phasing of the combustion with the movement of the piston, and knock becomes more likely. Figure 4.8 shows knock intensity, the maximum

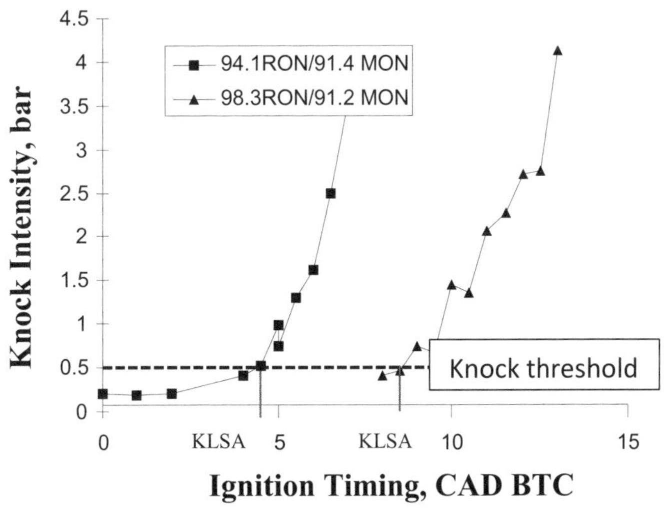

Fig. 4.8. Knock intensity increases as ignition timing is advanced. Data are from [4.39].

peak-to-peak amplitude of the filtered pressure signal averaged over 400 consecutive cycles, plotted against spark timing for two fuels.

The spark timing is shown as the ignition advance from TDC—the larger its value, the earlier the spark. The engine was running at 2000 rpm and wide-open throttle (further details about the fuels and the experimental conditions are in [4.39]). The spark advance when the knock intensity reaches the threshold value is known as the knock-limited spark advance (KLSA) and is commonly used as a measure of the knock resistance of a fuel in engine experiments. At the condition considered, the fuel with the higher RON of 98.3 has the higher knock resistance because it has a higher value of KLSA.

For a given fuel at a given operating condition, there is a positive correlation between knock intensity and (a) peak pressure [4.33, 4.40], (b) the difference between the peak pressure and the pressure at knock onset (i.e., the jump in pressure brought about by knock [4.33, 4.41]), and (c) the estimated maximum volumetric heat release rate resulting from knock [4.42]. The amplitude of the pressure fluctuation generated by autoignition increases with the propagation velocity of the autoignition reaction front emanating from the hot spot relative to the unburned gas u_a [4.43, 4.45]; see Section 5.2 for further discussion. However, if u_a is high enough, the local overpressure will be high enough to induce further chemical reaction at the reaction front. The pressure wave and the chemical reaction become coupled, and the reaction front takes on the characteristics of a developing detonation. The space inside the cylinder is too small for the reaction front to become a fully developed detonation wave where u_a reaches the local acoustic velocity, but the pressure pulse can become very large [4.43–4.45].

4.2.2 Knock-Limited Performance

Figure 4.9 shows brake torque plotted against ignition timing corresponding to the data in Fig. 4.8. The torque is maximum at a given speed and throttle setting at MBT (minimum advance for best torque) timing, which is 12 crank angle degrees before top center (CAD BTC) for the condition considered.

The ignition timing cannot be advanced beyond KLSA because of knock. If KLSA is lower than MBT, the engine is said to be knock-limited at that condition, because it cannot attain maximum torque. The torque at KLSA is the knock-limited torque (KLT). In Fig. 4.9, the 94.1 RON fuel is clearly knock-limited because it cannot be run near the MBT timing without running into high knock intensities. Most modern high-performance engines are knock-limited in at least some part of their operating regime on normally available fuels. Such engines are usually equipped with knock sensors. With the engine operating on a given fuel, when knock is detected, the engine management

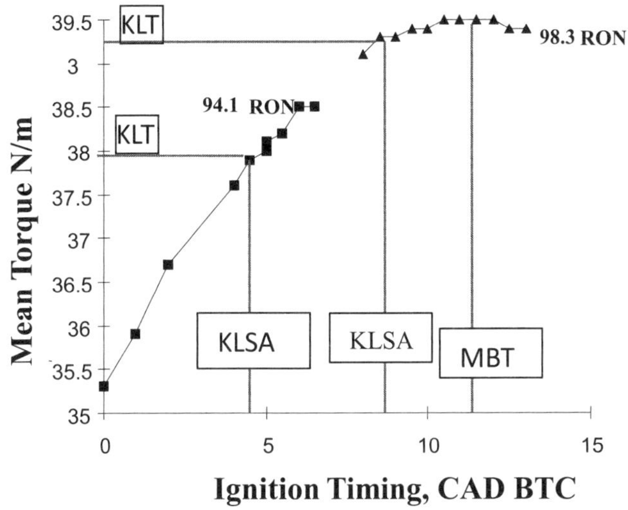

Fig. 4.9. Mean torque versus ignition timing corresponding to knock intensity results in Fig. 4.8. The 94.1 RON fuel is more knock-limited and has to be run further away from MBT timing compared to the 98.3 RON fuel.

system takes corrective action, usually by retarding the ignition timing, which causes a reduction in power [4.46–4.51]. In some turbocharged engines, such corrective strategies might also include the reduction of the boost pressure [4.47]. The engine will normally be set to operate at the optimum ignition timing. Typically, if knock is detected, ignition timing is usually retarded slightly, and if knock is still detected a fixed time later, the timing will be further retarded. However, if knock is *not* detected for a set length of time, the timing will be advanced in steps until the optimum timing is again reached. Thus, the engine is operated at the optimum ignition timing as often as possible. In such vehicles, if the antiknock quality of the fuel is increased, the chances of knock are reduced, and the engine is more likely to operate near the optimum condition, so performance can be improved up to a limit and either the constant speed power or an acceleration time can be used as measures of the antiknock quality of the fuel [4.15, 4.46–4.51].

4.2.3 Antiknock Quality of Practical Fuels—The Octane Index and the K Value

Engines can be run at different pressures and temperatures depending on the intake conditions and engine design. Figure 4.10 shows temperature in the unburned gas plotted against the pressure for the RON and MON tests based

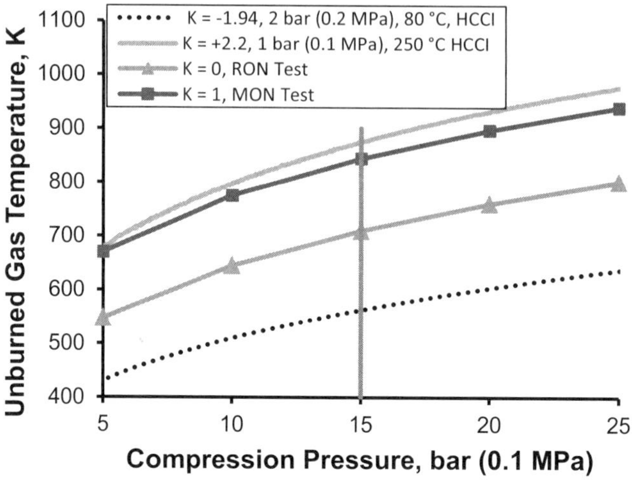

Fig. 4.10. Unburned gas temperature versus pressure at different engine conditions.

on the results from [4.25] and for the two HCCI test conditions considered in Fig. 4.5.

Modern SI engines are "beyond RON" because the temperature of the unburned mixture, at a given pressure, is even lower than in the RON test. Alternatively, the pressure for given temperature is higher in modern engines compared to the RON test. This is a consequence of the higher efficiency of modern engines. Engine designers have always tried to force more air into the cylinder while minimizing the temperature to improve power density and efficiency. Indeed, all the approaches aimed at improving the efficiency of SI engines such as increasing the compression ratio, direct injection, and turbocharging aligned with down-sizing will push modern engines further beyond RON. If, for a given pressure, the unburned gas temperature is *larger* than in the MON test condition, the operating condition can be said to be beyond MON. HCCI engines can be run well beyond RON or MON, and experiments in HCCI engines have helped in the understanding of fuel effects on autoignition as engine operating conditions change.

As discussed in Section 4.1.2, the same non-PRF can match different PRFs for autoignition at different pressure and temperature regimes. It has long been known that RON or MON alone does not adequately describe the knocking behavior of a practical fuel in an SI engine [4.30, 4.52–4.55]. In general, if z is a measurement parameter that depends on the autoignition quality of the fuel in an engine or a vehicle test, it is possible to find a linear regression equation [4.56]:

$$z = a\text{RON} + b\text{MON} + c \tag{4.4}$$

z can be the KLSA in engine tests; either an acceleration time or knock-limited power (KLP) in vehicles equipped with knock sensors or CA50, the crank angle at which 50% of the total heat release from autoignition, occurs in HCCI tests. See Section 4.3 for further discussion of HCCI measurements.

The octane index *OI* is then defined with RON and MON and the coefficients from the linear regression:

$$OI = \frac{a}{a+b}\text{RON} + \frac{b}{a+b}\text{MON} \tag{4.5}$$

Eq. 4.4 can then be rewritten with the octane index:

$$z = c + (a+b)OI \tag{4.6}$$

We now introduce a constant *K* that is defined as:

$$K = \frac{b}{a+b} \tag{4.7}$$

OI can now be rewritten as:

$$OI = (1-K)\text{RON} + K\text{MON} = \text{RON} - KS \tag{4.8}$$

Here *S* is the sensitivity of the fuel, (RON – MON). The octane index of a fuel is the octane number of the PRF that matches it for autoignition behavior in the particular test under consideration. *K* is a constant depending on the pressure and temperature evolution in the unburned gas—it is not a primary property of the fuel. The higher the *OI*, the greater the resistance to autoignition, and in the context of SI engines, the better the antiknock quality of a fuel at the particular condition. For a PRF, since $S = 0$, the *OI* is the same as its RON or MON. By definition, for the RON test condition, $K = 0$ and OI = RON, and for the MON test condition, $K = 1$ and OI = MON. For a condition beyond RON, *K* is negative [4.39, 4.56].

For a practical fuel that is not a simple binary mixture of iso-octane and n-heptane, RON > MON, and *OI* depends on the value of *K* as well as on its RON and MON. The value of *K* varies with engine operating conditions, and hence the same practical fuel will have different *OI* and will match different PRF for autoignition qualities at different operating conditions. The *OI* combines information about the autoignition or knock behavior of the fuel at two very different conditions (RON and MON) with information about the actual operating condition in the form of *K*. The sensitivity *S* of a fuel is a measure of the difference in the fuel chemistry from that of a PRF, and the *K* value is a measure of how different the pressure and temperature variation is from the RON test condition—if the test condition is below the RON test condition in Fig. 4.10, *K* is negative, and above the RON test condition, *K* is positive.

Modern SI engines have negative values of *K*—for a given RON in these engines, a lower MON fuel has higher *OI* and thus is more resistant to knock [4.15, 4.39, 4.46, 4.47, 4.56–4.62]. This is illustrated in Fig. 4.11 using data from [4.15] for a DISI engine with a compression ratio of 12.5 running full throttle at 2000 rpm. KLSA was determined for 11 fuels of different chemical compositions. In such experiments it is important to ensure that the RON and MON are correlated as little as possible across the set of fuels used—the set should consist of PRF as well as non-PRFs of different chemical composition, so that the sensitivity varies for comparable RON. If RON and MON are correlated, as can happen when fuels of intermediate octane quality are made by blending two high- and low-octane fuels, the effects of RON and MON cannot be separated. Figure 4.11a shows that there is a reasonable correlation between KLSA and RON, but there is little correlation between KLSA and MON (Fig. 4.11b). Figure 4.11c shows that there is an excellent correlation between KLSA and *OI*, with $K = -0.41$. The value of *K* was determined by using KLSA for *z* in Eq. 4.4. Note, however, that the relationship between KLSA and *OI* in Fig. 4.11c is not linear, as in Eq. 4.6. An alternative way to establish the value of *K* is to select a nonlinear model for the relationship between KLSA and *OI* and find *K* by trial and error, by finding the best fit between the model and the observations. However, this approach is more laborious and does not yield a value for *K* that is significantly different from the linear approach outlined above.

Chassis dynamometer tests in many modern knock-sensor equipped cars—using either an acceleration time or the power at the wheels at wide-open throttle at a constant speed as a measure of autoignition quality *z* in Eq. 4.4—have demonstrated that modern vehicles do indeed have very negative values of *K*. For a given RON, higher MON is detrimental to the antiknock quality of the fuel [4.15, 4.46, 4.47, 4.56]. In engine experiments, the value of *K* increases with increasing intake air temperature [4.57, 4.58] and decreases with increasing intake pressure [4.57, 4.61, 4.62]. Also, *K* becomes more negative as the compression ratio is increased [4.15, 4.39, 4.57, 4.59]. These observations are consistent with the fact that in Fig. 4.10, *K* decreases from 1 at the MON test condition to 0 at the RON test condition and becomes negative when the operating condition moves beyond RON [4.15, 4.39, 4.56]. *K* also decreases as non-PRFs with sensitivity $S > 0$ become relatively more resistant to autoignition as the pressure is increased for a given temperature (as discussed in Section 4.1.2). Tests in several engines show that as engine speed increases, with all

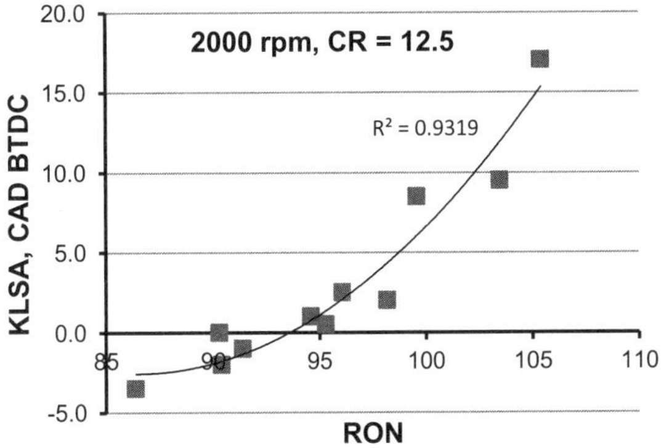

Fig. 4.11a. KLSA versus RON. Single-cylinder DISI engine running at wide-open throttle and stoichiometric mixture strength. Data are from [4.15].

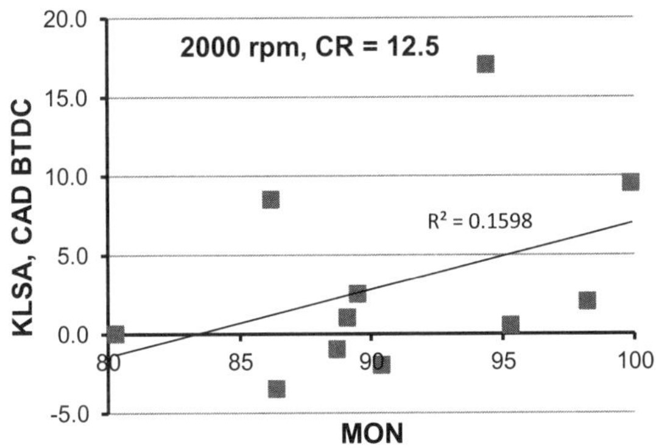

Fig. 4.11b. KLSA versus MON. Data are from [4.15].

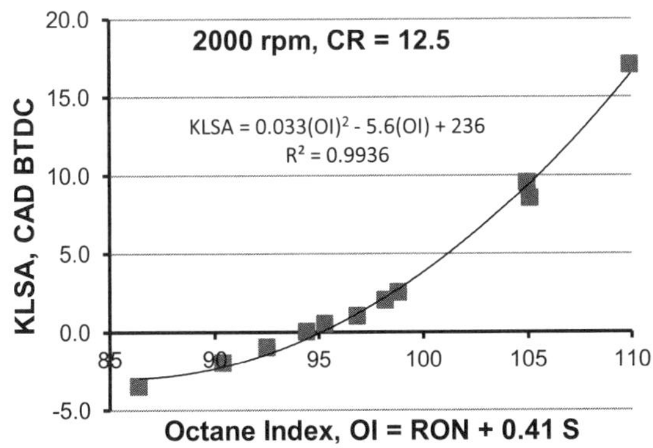

Fig. 4.11c. KLSA versus OI with $K = -0.41$. Data are from [4.15].

else fixed, the value of *K* increases [4.15, 4.39, 4.57–4.59], though this trend is not always observed [4.58, 4.61]. The increase in *K* with increasing engine speed might simply be reflecting the fact that as engine speed increases, heat transfer to the coolant decreases and in-cylinder temperatures increase. Thus, at low speeds, the operating condition could be beyond RON with a negative *K* value. However, as the engine speed increases, the temperature in the unburned gas at a given pressure increases, and the operating condition could cross the RON test condition in Fig. 4.10 and the value of *K* could become positive. The value of *K* also depends on the mixture strength, as will be discussed in Section 4.3. However, in SI engines this effect is not relevant because, in practice, SI engines have to run at or around stoichiometric conditions.

The RON test was invented in 1930 when the average compression ratio was around 5 [4.63], and engines had poor efficiency compared to today's engines. It was soon obvious that the RON test did not predict the antiknock quality of fuels on the road at that time. The MON test was then developed, and it reflected the antiknock behavior of fuels in cars of that time much better. Clearly, the *K* value of the U.S. car fleet then was around 1. The Coordinating Research Council in the United States has conducted frequent surveys to establish the octane appetite of U.S. cars between 1947 and 1996. Based on these data and other available sources, Mittal and Heywood [4.63] showed that the *K* value has decreased from around 1 to 0 in 2008 because of reduced end-gas temperatures and higher pressures. A decreasing trend for *K* in the U.K. car fleet has also been shown in [4.56]. In the United States, fuel antiknock quality is assumed to be best described by (RON + MON)/2, and *K* is assumed to be 0.5, probably because the U.S. car fleet used to have such an octane appetite when this specification was first agreed upon. However, engine technology has moved on, and as engine efficiency and power density have improved, the pressure and temperature regime has moved from the MON test condition in Fig. 4.10 toward and then beyond the RON test condition. On average, the *K* value has moved toward zero and then beyond to negative values in modern European and Japanese cars [4.15, 4.46, 4.47, 4.56]. Further developments in SI engine technology such as downsizing and turbocharging can only accentuate this trend [4.61,4.62]. The implications of this for future fuel specifications and manufacture will be discussed in Chapter 7.

4.2.4 Octane Requirement

At a given operating condition, the octane requirement (*OR*) of the engine is the *OI* of the fuel with KLSA the same as MBT. Such a fuel, or a fuel with *OI* higher than the *OR*, allows the engine to run at the best efficiency and power without knock. Figure 4.12 shows the knock-limited torque—the torque at KLSA—plotted against KLSA for different fuels for the condition considered in Fig. 4.11. The last point on the curve, with a KLSA of 17 CAD enables the

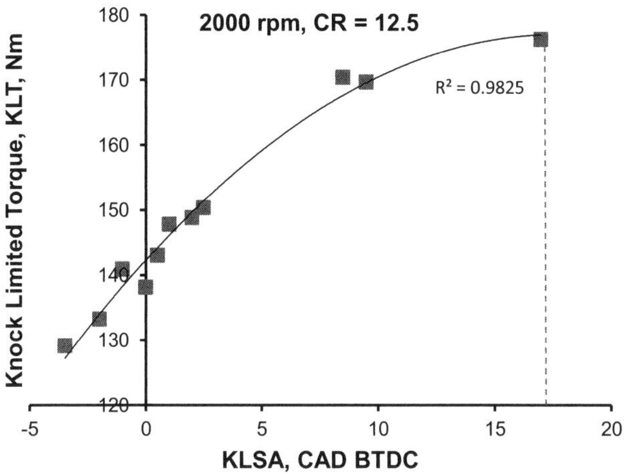

Fig. 4.12. Knock-limited torque (KLT) versus KLSA corresponding to data in Fig. 4.11 taken from [4.15].

engine to attain the highest possible torque at this operating condition without being limited by knock. From Fig. 4.11c, this fuel has an *OI* of 113, and hence the *OR* of the engine at this operating condition is 113.

In vehicles equipped with knock sensors, at each speed and load there will be a default setting of ignition timing to which the engine is supposed to return even if it is temporarily moved away from it because of knock. The *OR* of such a vehicle is the minimum *OI* that allows this default value to be reached. Usually, this default setting is less advanced than MBT timing because of safety margins to account for uncertainties in the knock-sensor operation and other concerns associated with emissions control. Therefore, the *OR* of the vehicle is determined by the control strategy employed by the engine management system and is often lower than the real *OR* of the engine, which is associated with the best performance the engine is capable of at MBT timing.

Ideally, the *OI* must equal or exceed *OR* at all operating conditions. For instance, as engine speed increases, *OR* decreases, knock becomes less likely because of the reduced time available for autoignition, but *K* increases in most cases [4.15, 4.39, 4.57–4.59]. In fact, there is an empirical trend of decreasing *K* with increasing *OR*, as illustrated in Fig. 4.13 for different engine experiments from [4.15] and [4.39].

This simply reflects the fact that in a given engine an increase in *OR* by increasing compression ratio or intake pressure, or in some cases, reducing speed, will move the pressure and temperature regime downwards in Fig. 4.10—toward and beyond RON. The consequences of such changes have been explored for the engine studied in [4.15]. If two fuels have the same *OI* at a

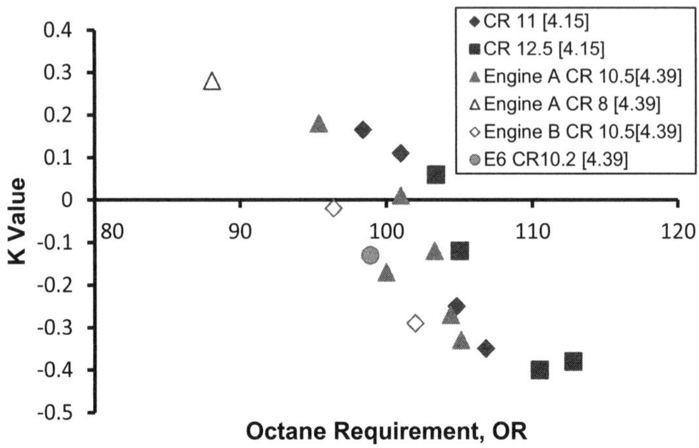

Fig. 4.13. There is a general trend for a decrease in *K* value with increasing octane requirement (*OR*).

condition when *K* is positive at a lower *OR*, the *OI* of a fuel of higher sensitivity will be nearer to satisfying the *OR* of the engine at the higher *OR* condition than a fuel of lower sensitivity [4.15]. This can be illustrated as follows. Let us consider two fuels: Fuel F1 has a research octane number of RON1 and sensitivity *S*1; Fuel F2 has a research octane number of RON2 and sensitivity *S*2. Let us assume *S*1 > *S*2 and that the *OI* of the two fuels equals the octane requirement, *ORx* at an operating condition when *K* is positive, say, + *K*1, that is, *ORx* = RON1 – *K*1 ∗ *S*1 = RON2 – *K*1 ∗ *S*2. Hence, *N*2 = RON1 – *K*1 ∗ (*S*1 – *S*2). At an operating condition when *K* is negative, say, –*K*2, let us assume that the *OI*, (RON1 + *K*2 ∗ *S*1) of Fuel F1 equals the new, higher octane requirement *ORy*. The *OI* of the less sensitive fuel, Fuel F2 will be (RON2 + *K*2 ∗ *S*2) and will fall short of required *OI*, *ORy* by (*K*1 + *K*2)(*S*1 – *S*2). Thus, a more sensitive fuel, Fuel F1, will be able to satisfy the octane requirement of the engine over a wider operating range.

There is an interesting curiosity: If a linear relationship between *K* and *OR* were to exist,

$$K = A - B[OR] \tag{4.9}$$

We would also want *OI* = *OR* and from Eqs. 4.8 and 4.9:

$$\mathrm{RON} - AS = [OR][1 - BS] \tag{4.10}$$

This identity will be satisfied always if $S = 1/B$ and RON = A/B; a fuel with these properties would satisfy the engine's requirement at all conditions.

At a given operating condition, for fixed ignition timing, the pressure and torque obtained will also depend on the flame development rate—a faster flame will usually lead to higher pressures in the cylinder, which makes autoignition more likely. Faster flame development also reduces the time available for autoignition. Engine combustion is also marked by cyclic variations, and for a given fuel, the pressure effects dominate, and in general, faster burning cycles are more prone to knock [4.33]. In the discussion above, as also in knock tests like the RON or MON tests, the effect of differences in burning rates between different fuels was not considered. Similarly, the effect of the differences in charge cooling brought about by the evaporation of different liquid fuels was also not considered; this is particularly relevant in DISI engines. Of course, charge cooling also leads to higher volumetric efficiency and hence higher pressure if the fuel is injected when the intake valve is still open, which might counteract the effect of reduced temperature. Such differences between fuels do exist, but are small compared to the differences in autoignition chemistry, which can be reasonably assumed to be the main source of difference in the knock behavior of practical fuels.

4.3 Fuel Effects on Homogeneous Charge Compression Ignition Combustion

HCCI engines can be run at very lean mixture strengths where NOx and smoke can be extremely low. SI engines cannot be run at such conditions because a stable flame cannot be maintained; in diesel engines, the locally rich mixtures where NOx and smoke are generated cannot be avoided, even though the overall mixture is kept very lean. HCCI engines can also be run without throttling and, compared to SI engines, at much higher compression ratios so that their efficiency is higher. Preliminary studies in HCCI in the 1980s [4.64, 4.65] were followed by a new surge in interest starting in the late 1990s [4.66–4.68]. A review of the literature can be found in [4.69].

However, HCCI engines have very low power density—they have to be run at lean conditions to avoid excessive and damaging maximum heat release rates and to maintain low NOx levels. They are very difficult to control since the combustion phasing can only be controlled by initial conditions that determine the autoignition behavior of the mixture and the combustion phasing. The lean and low-temperature combustion required to keep the level of particulates and NOx low in HCCI engines inevitably leads to high CO and HC emissions, and these are difficult to control by aftertreatment because of low exhaust temperatures. It has now become apparent that it is not possible to have a practical and affordable engine operating over its required speed/load range in pure HCCI mode while meeting the requirements on combustion stability, emissions, and maximum pressure rise rates. In this context, by HCCI we mean compression ignition where fuel and air are *fully* premixed. As discussed in Chapter 6, even when fuel and air are not fully but are sufficiently mixed by relatively late injection of a fuel that is relatively resistant to autoignition, many of the advantages of HCCI combustion (i.e., low NOx, low smoke, and high efficiency) can be obtained, but with in-cycle control over combustion phasing. Sometimes this type of partially premixed compression ignition is confusingly referred to as HCCI in the diesel literature. Of course, even in HCCI engines, even though the fuel and air are fully premixed by injection of the fuel very early in the engine cycle, the mixture is never truly homogeneous because of temperature and mixture strength gradients [4.70, 4.71], and autoignition always starts at localized hot spots [4.72, 4.73].

Research into HCCI combustion has helped to advance our understanding of autoignition, particularly with respect to fuel effects, because HCCI engines can be run over much wider operating ranges of pressure, temperature, and mixture strength compared to knocking engines. There have been many studies where different fuels of different chemical composition have been tested in the same HCCI engine [4.2, 4.28, 4.29, 4.74–4.83]. In most of these studies, various fuels are tested in the engine at a fixed engine speed, mixture strength, EGR level, and intake temperature and pressure—all the mixtures follow a single polytropic line during compression such as those shown in Fig. 4.10. Depending on their autoignition behavior, different fuels start deviating from the compression polytrope at different times as autoignition reactions set in. Fuel autoignition quality can be characterized by a combustion phasing parameter such as CA50, the crank angle at which 50% of the total heat release occurs. The higher the value of CA50 (i.e., the later the heat release), the more resistant the fuel is to autoignition. In all these experiments, as expected, different fuels have different combustion phasing, but this does not necessarily correlate with RON or MON. However, an approach based on the *OI* provides a general framework within which the autoignition of a wide range of fuels exposed to a wide range of environments in the HCCI engine can be understood. The group in Sandia National Laboratories has taken a different approach to ranking fuels [4.83] in HCCI engines. In their experiments, with all else fixed, the intake temperature is changed to ensure that CA50 occurs at TDC—the more resistant the fuel is to autoignition, the higher the required intake temperature.

4.3.1 Understanding Fuel Effects in HCCI Engines within a Framework Based on the Octane Index

HCCI engines can be run well beyond RON or beyond MON conditions (see Section 4.2.3), so that there might not be any correlation between the actual measure of the autoignition resistance or quality of the fuel such as CA50 at the particular operating condition and either RON or MON. This is illustrated in Figs. 4.14a and 4.14b, where the average CA50 has been plotted against RON and MON, respectively, for HCCI experiments done in a single-cylinder engine running at 1200 rpm, with the intake pressure and temperature at 2 bar (0.2 MPa) absolute and 80 °C, respectively, and the normalized air/fuel ratio λ (the reciprocal of the equivalence ratio, Φ) at about 4.5. The engine was run on 11 different fuels, as described in [4.29], and the data are listed in Appendix 2 of [4.29]. The *K* value to be used in the *OI* can be found by using CA50 for *z* in Eqs. 4.4 and 4.7. In this particular case, the value of *K* is –1.94, and there is a good correlation between CA50 and OI = RON + 1.94S in Fig. 4.14c. The correlation might have been even better if the mixture strength and other operating conditions could have been controlled more tightly for all the fuels. Several illustrations of such good correlations between CA50 and OI, when no correlation exists between CA50 and RON or MON are given for negative values for *K* [4.2, 4.78, 4.84] as well as for cases $K > 1$, that is, with the engine running "beyond MON" [4.28].

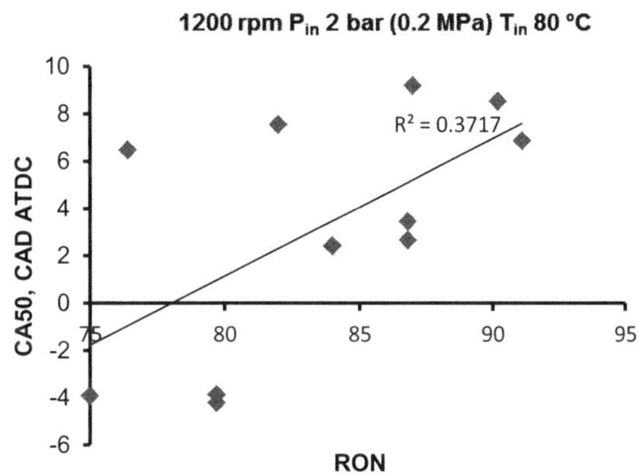

Fig. 4.14a. CA50 versus RON for an HCCI engine running at 1200 rpm, 2 bar (0.2 MPa) absolute intake pressure, 80 °C intake temperature, λ = 4.5 ± 0.1. Data are from Appendix 2 of [4.29]. Engine compression ratio is 14; for further details, see Ref. [4.29].

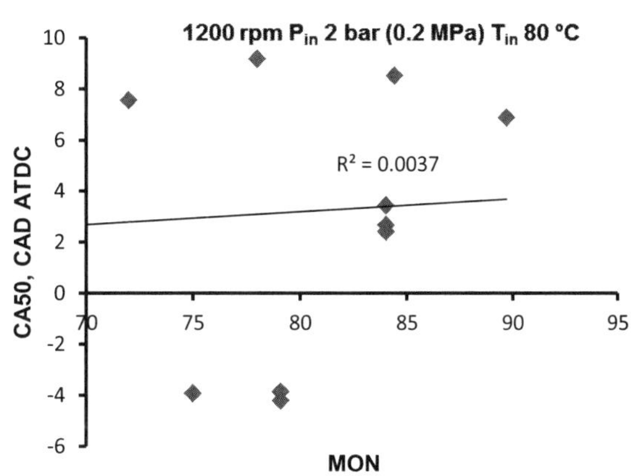

Fig. 4.14 b. CA50 from Fig. 4.14a plotted against MON.

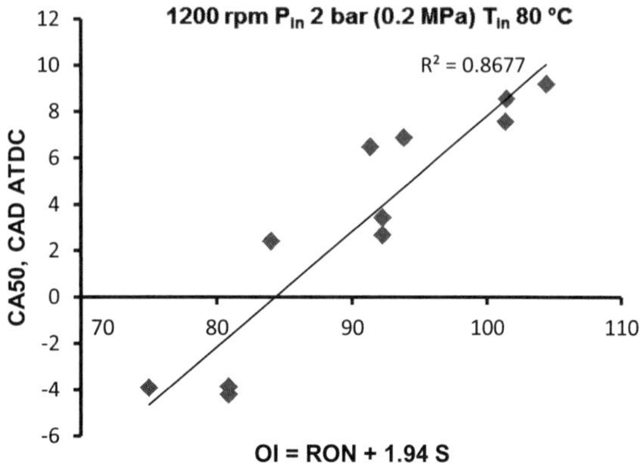

Fig. 4.14c. CA50 from Fig. 4.14a plotted against $OI = 1.94RON - 0.94MON = RON + 1.94S$.

The autoignition quality requirement of the engine at each condition OI_0 is defined as the octane index when CA50 occurs at TDC (i.e., CA50 is 0). From Eq. 4.6, with z = CA50:

$$OI_0 = \frac{-c}{(a+b)} \tag{4.11}$$

We can now rewrite Eq. 4.10 as:

$$\text{CA50} = (a+b)(OI - OI_0) \tag{4.12}$$

Thus when the fuel autoignition quality OI is equal to the engine's requirement OI_0 CA50 will be 0 (i.e., at TDC).

Combustion in HCCI engines depends strongly on the temperature/pressure variation in the combustion chamber. The temperature in the unburned gas, when the pressure is 15 bar (1.5 MPa), T_{comp15}, was introduced in [4.2] to mark the position of the temperature/pressure polytrope such as those shown in Fig. 4.10 for any given experiment; the value of 15 bar (1.5 MPa) chosen for the pressure has no special significance. By changing inlet conditions, internal and external EGR levels, compression ratios, and mixture strengths, two different HCCI engines have been run along many different T versus P characteristic lines. For instance, initial mixture temperature at the start of compression can be increased either by heating the inlet air or by using a lot of hot internal EGR created by negative valve overlap; initial pressure can be increased by turbocharging the engine. The consolidated results from the two engines are considered in [4.29] and are also discussed in [4.84].

4.3.1.1 Variation of CA50 with Octane Index

In several HCCI experiments with high in-cylinder temperatures, fuels of widely varying octane numbers were found to release heat roughly at the same time in the engine cycle [4.28, 4.80, 4.82]. In other words, regarding the sensitivity of CA50 to the *OI*, the slope $(a + b)$ of the line was very small (e.g., Fig. 4.14c). In Fig. 4.15, $(a + b)$ is plotted against T_{comp15} for a wide range of HCCI experiments in two very different HCCI engines; these data are from Fig. 5 of [4.29].

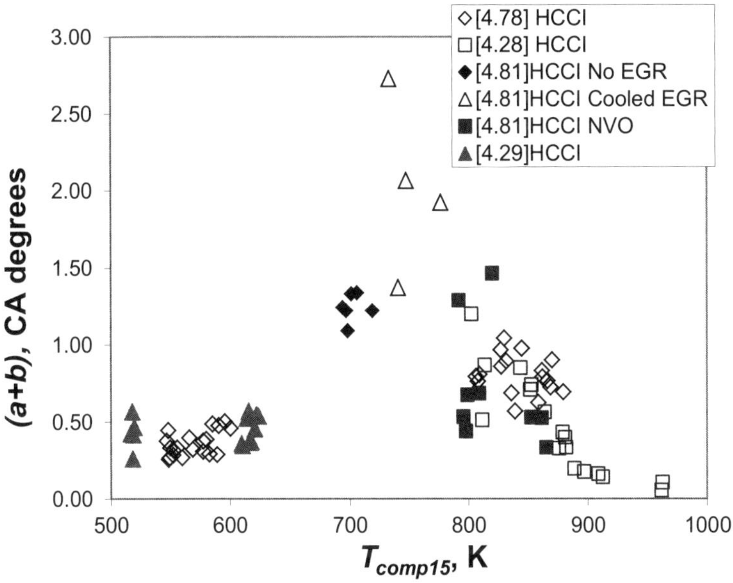

Fig. 4.15. The slope $(a + b)$ of CA50 versus *OI* plotted against T_{comp15}. At high and low values of T_{comp15}, the difference in combustion phasing between fuels of different *OI* is small.

One of these engines had a cylinder displacement of 1.95 L and was based on a heavy-duty diesel engine, while the other had a cylinder displacement of 0.5 L and was based on a gasoline engine. Each of the 88 experimental points in Fig. 4.15 was derived from testing different fuels with the engine conditions such as speed, equivalence ratio, intake pressure, intake temperature, and EGR levels fixed.

At both high and low values of T_{comp15}, $(a + b)$ is small. T_{comp15} can be increased by increasing the intake temperature as in [4.28, 4.82] or by retaining a lot of internal EGR gases by negative valve overlap [4.80, 4.81]. The differences in autoignition chemistry between different fuels decrease as temperature increases. For instance, the difference in ignition delays scaled to a common pressure is small at high temperatures for different fuels (e.g., Fig. 5 in [4.20]); see also Fig. 4.6. Hence, if the in-cylinder temperature is high during an HCCI experiment, all the fuels will become very prone to autoignition, and the difference between fuels will be small. T_{comp15} is reduced by increasing the intake pressure, and autoignition becomes more likely (i.e., ignition delays are reduced) as pressure increases, even if the temperature is fixed (Eq. 4.2), and again the difference between different fuels decreases. Clearly, at extreme conditions, when $(a + b)$ is small, the absolute value of K will be large (Eq. 4.7), and attempts, including the one based on OI and K, to describe fuel differences become irrelevant since all fuels behave similarly within experimental accuracy.

4.3.1.2 Variation of K

In HCCI combustion, K varies widely but depends strongly on T_{comp15} (see Fig. 4.16—the data are from the same set of experiments considered in Fig. 4.15).

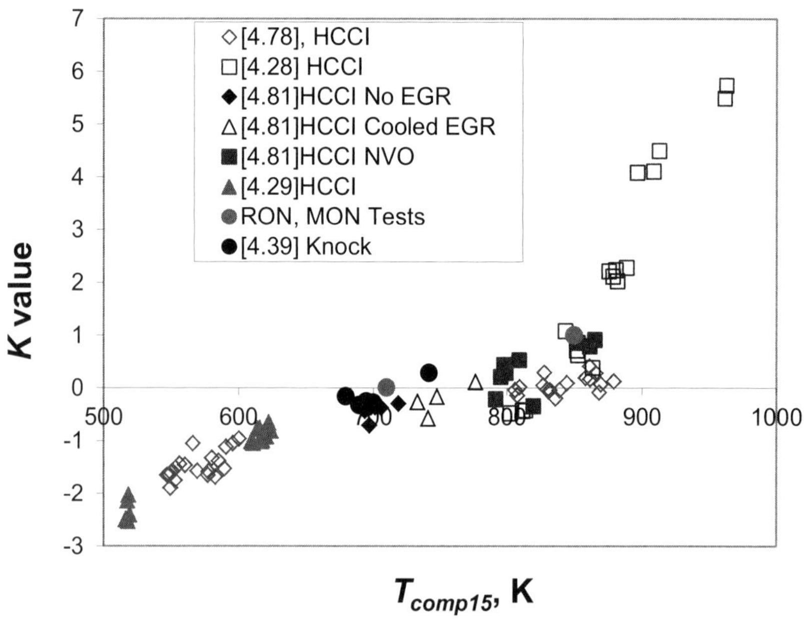

Fig. 4.16. Measured K values plotted against T_{comp15} for HCCI experiments and knock experiments. HCCI results are summarized in [4.29, 4.84]. Knock results are from [4.39].

Also shown are results from knock experiments in [4.39] and RON and MON tests; the results at only 1200 rpm are considered from [4.39] since estimates of T_{comp15} were available for them. There is also clear evidence, both from HCCI experiments [4.78] and knock experiments [4.58, 4.59], that K decreases as λ

increases, though Fig. 4.16 does not take this into account. An approximation for this trend for $\lambda > 2.5$ and $T_{comp15} < 825$ K in HCCI engines is given in [4.78] by:

$$K = 0.00497T_{comp15} - 3.67 - 0.135\lambda \tag{4.13}$$

The dependence on λ is not relevant to SI engines, which mostly run at $\lambda = 1$. Given the scatter in the data, Eq. 4.13 predicts the trend rather than the exact value of K. If T_{comp15} is greater than for the RON test K is positive, the *OI* of a non-PRF with $S > 0$ will be less than its RON. Leppard, in his classic paper [4.25], ascribed this to the fact that, compared to a PRF, non-PRFs are less likely to display NTC behavior where ignition delay increases with increasing temperature. For illustration, consider PRF 95 (95% volume iso-octane + 5% volume n-heptane) and a gasoline of 95 RON and 85 MON. They are matched for autoignition behavior at the RON test condition (i.e., they have the same RON). As the temperature in the unburned gas is increased, the mixture will spend more time in the region where NTC behavior occurs during compression before the final autoignition, and the PRF will become progressively more resistant to autoignition compared to the non-PRF. T_{comp15} is lower for the RON condition compared to the MON condition (Fig. 4.10). If T_{comp15} is raised to the MON test condition, the gasoline will be significantly more prone to autoignition compared to PRF 95 and will actually behave like PRF 85.

If T_{comp15} is reduced below the RON test condition, the pressure in the unburned gas for a given temperature will be higher. Under such circumstances, the non-PRF, with RON > MON, will become relatively more resistant to autoignition compared to PRF, as discussed in Section 4.1.2 and its OI will be higher than its RON (i.e., K will become negative). It is not possible to clearly decouple effects of pressure and temperature. The differences between PRF that define the RON and MON scales and practical fuels arise because of the different dependences of their autoignition chemistry and thus the autoignition delay times, on both pressure and temperature. Such differences have been explored in HCCI experiments in [4.85], which also discusses the relationship between K values and autoignition delay times.

4.3.1.3 The Octane Requirement OI_0 of an HCCI Engine

OI_0 at any particular condition is defined as the *OI* of the fuel that gives CA50 at TDC at that condition. It is experimentally determined by the linear model for the CA50 versus *OI* variation, such as the one shown in Fig. 4.14c. OI_0 also varies widely in HCCI engines but could be related through an empirical expression to the mixture's thermodynamic parameters and engine operating conditions [4.29, 4.84]:

$$OI_0 = 59 + 0.015T_{maxcomp}(\text{in } K) + 0.663P_{maxcomp}(\text{in bar abs}) - 2.6\lambda^* - 0.0123N(\text{rpm}) \quad (4.14)$$

Here,

$$\lambda^* = (1 + EGR_f)\lambda \quad (4.15)$$

Here, λ is the normalized air/fuel ratio, and EGR_f is the EGR mass fraction. EGR_f is defined to be 0 when the exhaust gases present in the mixture come only from residuals left in the cylinder at the end of the exhaust stroke. In [4.81] the effect of external EGR on HCCI combustion was studied, and in many of those cases when EGR_f was not 0. OI_0 is dimensionless, $T_{maxcomp}$ and $P_{maxcomp}$ are the compression temperature and pressure at TDC expressed in degrees Kelvin and bar (absolute), respectively. They depend on both the conditions at the start of compression and the compression ratio of the engine. N is the engine speed in rpm. Fig. 4.17 shows that there is reasonable agreement between the measured OI_0 and the value predicted by Eq. 4.14, even though the results are from two very different single-cylinder engines covering a very wide range of operating conditions.

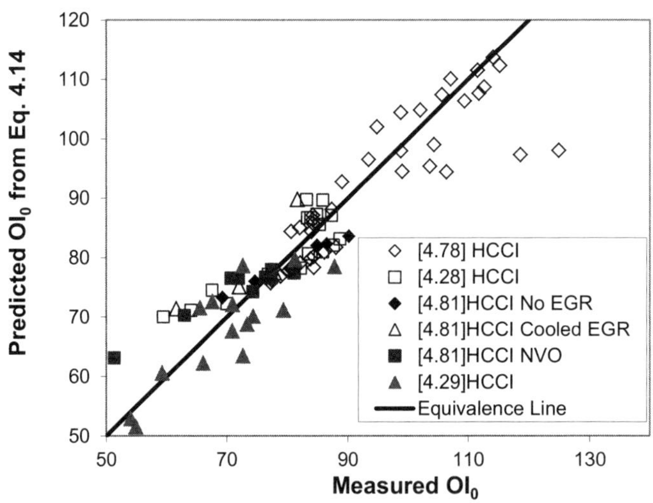

Fig. 4.17. OI_0 predicted by Eq. 4.14 plotted against the measured value.

In general, as indicated by Eq. 4.14, if either $T_{maxcomp}$ or $P_{maxcomp}$ are increased or λ^* or N is reduced, with everything else unchanged, we should expect autoignition to become more likely. OI_0 will be increased, which is to say that a higher OI fuel will be needed to counteract the increased likelihood of autoignition because of increased temperature/pressure or a richer mixture or a longer time available for autoignition.

Ideally *OI* must match OI_0 at all conditions. If the *OI* of the fuel exceeds OI_0, CA50 will be positive (i.e. will occur after TDC) in the engine cycle. If *OI* is increased for a fixed OI_0 or OI_0 is reduced because of changes in the engine operating conditions for a fixed fuel, heat release will occur later and later, and eventually autoignition will fail if *OI* exceeds OI_0 by too much. If, on the other hand, OI_0 is increased too far above the *OI* available from the fuel, heat release will be too early, the heat release rate and the pressure rise rate will be too high, and the engine will knock. Sensitive fuels have an advantage [4.2, 4.28, 4.78] in that their *OI* changes in the same direction as OI_0 as pressure and temperature in the engine change because of changes in *K*. For instance, for a given engine speed, IMEP (indicated mean effective pressure) increases linearly with fuel energy content in joules per cycle [4.28]. This in turn depends on the product of the air consumption rate and the equivalence ratio. Therefore, IMEP can be increased by boosting the intake in an HCCI engine for the same λ. Then, $P_{maxcomp}$ and $T_{maxcomp}$ will increase, and hence OI_0 will increase. Also, T_{comp15} will decrease, because the pressure will increase for a given temperature; *K* will decrease and could even become negative; and the $OI = \text{RON} - KS$ of a more sensitive fuel will be higher than that of a less sensitive fuel. Thus, for a given RON, a sensitive fuel will be more flexible and allow a greater range of operation in the HCCI engine.

4.3.1.4 Limits of Operation

There are two operating limits that are relevant to HCCI engine operation—the knock limit, when the pressure rise rate is unacceptably high, and the instability limit, when the cyclic variation is unacceptably high; these limits are not fixed objectively. The maximum pressure rise rate (MPRR) is related to the noise from the engine [4.79, 4.86], and the cyclic variation can be characterized by the COV (coefficient of variation = mean/standard deviation) of IMEP. With all else being equal, average MPRR in HCCI experiments increases, and the COV of IMEP decreases as the equivalence ratio increases (i.e., as the mixture is made richer) and as CA50 is reduced. As discussed in the previous section, CA50 depends on the difference between *OI* and OI_0. In addition, with all else being equal, MPRR increases with intake pressure and decreasing fuel sensitivity. Empirical correlations based on experiments linking average MPRR and COV of IMEP to these different parameters are presented in [4.29]. Also, in general HC and CO emissions increase as CA50 increases, and NOx emissions increase as CA50 decreases and MPRR increases.

An individual autoignition starts from a hot spot and the resulting overpressure above the ambient increases with the propagation velocity of the autoignition reaction front emanating from the hot spot relative to the unburned gas, u_a as discussed briefly in Section 4.2.1 and in more detail in Section 5.2. Bradley [4.45] has reviewed these matters.

4.3.1.5 Some Remarks about the Octane Index Approach

An empirical approach based on the *OI* and *K* value provides a suitable framework for understanding the experimentally observed behavior of fuels in the gasoline autoignition range (RON > 60, CN < 30) over a wide range of operating conditions. The *K* value can be related to T_{comp15}, which simply defines the pressure and temperature regime in the engine, and λ. T_{comp15} in turn is influenced by intake conditions, engine design, and speed. Engine design and operating parameters determine the pressure and temperature in the cylinder and, thus, T_{comp15}. Results from two very different engines line up with each other, suggesting that this approach, based on in-cylinder conditions, can be applied to different engines. The predictive value of this approach is limited by the scatter in the experimental results, but many of the trends are clearly identifiable. Operating parameters of interest like CA50 and MPRR can still be estimated using such empirical correlations if sufficient information to calculate T_{comp15} is known, operating conditions are specified, and the RON and MON of the fuel are known [4.84]. There are also some exceptions where some fuels do not follow the general trend of CA50 with *OI*; this has been observed when the in-cylinder temperature is high either because of high intake temperatures [4.28] or high internal EGR [4.81]. However, given the complexity of practical fuels and the state of knowledge about autoignition chemistry of such fuels, an empirical approach such as the one described gives the best chance of bringing order to a wide range of experimental results. It also helps to identify important areas for further research in fundamental chemical kinetics research (e.g., to understand pressure effects for different types of fuels).

4.3.2 Other Approaches to Characterizing Fuel Performance in HCCI Engines

Jeuland et al. [4.79] have introduced the controlled autoignition number (CAN) to enable the comparison of two fuels in an HCCI engine. The operating boundaries in terms of load and speed are first established in a given engine for a reference fuel. For instance, in Fig. 4.18, the reference fuel could be run satisfactorily only in the area bounded by the heavy lines—at a given speed, outside the boundary, and for high loads, the noise would be unacceptable; for low loads, the combustion would be too unstable. A similar operational regime is established for the test fuel: the area inside the dashed line in Fig. 4.18.

CAN is then defined as the ratio of the operational area of the test fuel to the operational area of the reference fuel. However, since autoignition behavior changes with pressure and temperature, the CAN ranking of a fuel will depend on the reference fuel chosen as well as on the engine operating

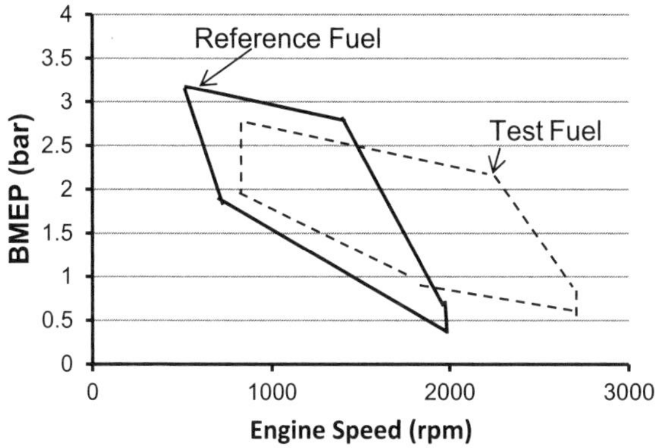

Fig. 4.18. Operating range of a HCCI engine on two different fuels.

conditions. It is suggested in [4.79] that there could be four CAN ratings for each fuel associated with low/low, low/high, high/low, and high/high speed/load conditions.

Shibata, Urushihara, and colleagues have used the LTHR and high-temperature heat release (HTHR) to characterize and explain the differences in autoignition behavior of fuels in HCCI experiments [4.82, 4.87–4.89]. They have demonstrated that n-paraffins display high LTHR, and aromatics and some naphthenes and olefins inhibit LTHR, and that these effects can be explained from chemical kinetics [4.89]. They have also related RON, MON, and the OI to HTHR and LTHR characteristics and demonstrated that the K value reflects the trade-off between LTHR and LTHR inhibitor effect in each test condition. They then propose an HCCI index, which is a linear function of MON and volume concentrations in the fuel of n-paraffins, iso-paraffins, olefins, aromatics, and oxygenates. The coefficient of each of these parameters is a function of pressure and temperature in the cylinder. Thus, this HCCI index, like OI, will depend on the fuel composition as well as engine operating conditions.

Ryan and colleagues have used the elevated pressure autoignition temperature, EPAIT, based on measurements of ignition delay in the ignition quality test (IQT) (see Section 2.2.2) at different temperatures, to characterize fuel quality of diesel fuels in HCCI engines [4.90, 4.91]. EPAIT is the temperature that yields a fixed ignition delay and correlates almost linearly, as does the ignition delay at a fixed temperature in the IQT, with the cetane number for fuels in the diesel autoignition range ($CN > 30$). In fact, in HCCI engines, the derived cetane number (DCN) correlated well with CA50 for fuels of different chemical composition in the diesel autoignition range [4.92].

In general, definition of autoignition quality is simpler for practical diesel fuels than for gasoline fuels [4.56]. For instance, if two diesel fuels of different cetane numbers are mixed, the cetane number of the blend can be predicted well by a linear formula that is the sum of the individual cetane numbers multiplied by the volume ratio. However, RON or MON of a mixture of fuels of different molecular structure cannot be usually predicted by linear functions [4.6]. The autoignition quality of a diesel fuel can be adequately described by a single quantity measured at one experimental condition such as the ignition delay in the IQT. The ranking of autoignition propensity for fuels of different chemical compositions in the diesel autoignition range obtained in the IQT will remain substantially the same if the same fuels are tested at very different conditions in a diesel engine.

However, a single ignition delay measurement cannot describe the autoignition behavior of fuels in the gasoline autoignition range adequately, as discussed in Section 4.1. A range of pure alkanes and PRFs were tested in a rapid compression machine in [4.93]. Even for these paraffinic fuels, the correlation between RON and an ignition delay measured at one temperature was poor. Ignition delay was also measured for different fuels in a rapid compression machine in [4.92]. Figure 13 in [4.92] shows that for fuels of similar composition, such as PRFs, there was a correlation between ignition delay and RON. However, for comparable RON, there was a wide variation in ignition delay for different chemical classes.

For practical gasoline fuels, at least three pieces of information are needed to characterize the autoignition quality of fuels in the gasoline autoignition range—two measures such as RON and MON, which describe the autoignition behavior at two very different pressure/temperature evolutions with time, and a third parameter such as K in OI, which describes the actual engine operating condition in relation to the specification test conditions. In the HCCI index proposed in [4.89], the volumetric fuel composition information and the coefficients of the parameters incorporate the information reflected by RON and K in OI. The autoignition behavior of an aromatic component in a diesel fuel is essentially determined by the long chain attached to the aromatic ring. In a gasoline aromatic component, the length of the chain attached to the aromatic ring is much shorter and the aromatic ring has a significant influence on autoignition chemistry. This might be one of the reasons why it is easier to define the autoignition quality of practical diesel fuels in comparison to practical gasoline fuels.

4.3.3 Fuel Requirements of HCCI Engines

HCCI combustion is again taken to mean *fully* premixed compression ignition. The fuel must be volatile enough to allow sufficient premixing of the fuel with air. Otherwise there will be fuel-rich regions that will cause an increase in NOx and soot, defeating the primary purpose of HCCI combustion. The deficiency of a fuel in this respect can be overcome to a great extent by better mixture-preparation strategies using direct injection, improved injector design, high injection pressures, flow control, and so on. Nevertheless, a more volatile fuel will have an advantage over a less volatile fuel in HCCI engines. The autoignition quality of the fuel must satisfy the changing requirements of the engine—*OI* must equal OI_0—over as wide a range of operating conditions as possible. At high loads, with low l and high $P_{maxcomp}$ and $T_{maxcomp}$, OI_0 will be high, and hence *OI* needs to be high; though at low loads the failure to autoignite will be a problem and *OI* needs to be low. This is usually very difficult with a single fuel since OI_0 changes so widely, but as discussed earlier, sensitive fuels have an advantage in this respect. Again, engine control can be employed to overcome deficiencies of the fuel and enable the engine to continue running, though not necessarily at the best possible efficiency. For instance, if *OI* is too large compared to OI_0, and autoignition fails, charge temperature could be increased sufficiently by retaining large amounts of internal EGR. If the temperature is high enough—in the region where $(a + b)$ tends to zero in Fig. 4.15—autoignition could be ensured with most fuels.

4.3.3.1 Idealized "Full HCCI" Scenario

We first discuss the idealized, but very unlikely, scenario where the engine will run in HCCI mode over its entire operating regime. The fuel classes considered in this section have been described in Chapter 2.

Gasoline fuels have an advantage over conventional diesel fuels because they are more volatile. Conventional diesels are also very prone to autoignition with *OI* less than about 35. OI_0 with a compression ratio of 14 is around 50 at low loads and needs to be much higher at high loads when the equivalence ratio is higher and λ is lower. Thus, it will be much more difficult to attain satisfactory HCCI combustion with diesel fuels unless OI_0 is reduced by reducing the compression ratio ($P_{maxcomp}$ and $T_{maxcomp}$ in Eq. 4.14). With all things being equal, a fuel with a higher *OI* will enable a higher maximum load in HCCI operation. GTL fuels are in general even more prone to autoignition than conventional diesels, though they have higher volatility. Kerosenes would be better than conventional diesels because of better volatility and lower cetane quality. On the other hand, there could be difficulties in ensuring that gasoline fuels can autoignite at low loads because of their high *OI*. However, as already discussed, this deficiency could be more easily overcome by control strategies such as retaining high levels of hot internal EGR or other thermal management.

Gasoline fuels can be blended to have high sensitivity, which gives them more flexibility in HCCI engines. In this respect, they will be better than naphtha, which will have the right volatility but will be less sensitive, and it is possible that its *OI* will be lower than desired. In summary, conventional gasoline fuels are more suited for a full HCCI engine than conventional diesel fuels. They need to be as sensitive as possible and have RON values probably at the levels of current gasoline fuels. Of course, these judgments are relative in that an HCCI engine can be run on any fuel if the conditions are chosen to suit it. From the point of view of extended operating range to higher loads, ease of mixture formation, and flexible ignition quality (which might make engine control a little easier), fuels can be ranked from the best to the worst as follows: gasoline, naphtha, kerosene, appropriate GTL cuts, and conventional diesel.

4.3.3.2 Practical Scenario

In practice, HCCI engines have to be dual-mode engines—running in HCCI mode where possible and reverting to conventional diesel or SI mode where they cannot, namely, at startup and high load. In such engines the fuel will have to support the conventional mode of combustion. HCCI/SI engines, some prototypes of which have been developed, switch to SI mode at high loads, and their fuel requirement will be determined by the necessity to avoid knock in the SI mode at high loads. Such engines have higher compression ratios compared to current SI engines in order to extend the HCCI operating range at low load on the gasoline fuel. Therefore, ideally, they will require a fuel with high RON and high sensitivity.

At compression ratios used in current diesel engines, using current diesel fuels, it is very difficult to get HCCI combustion except at extremely low loads. Hence, a true HCCI/diesel engine is not a practical proposition. However, the main thrust of diesel engine development is toward attaining more premixed, HCCI-like combustion to reduce NOx and smoke. It has also become clear that the operating range of HCCI engines can be extended by stratification of the mixture [e.g., 4.93–4.97]. In other words, fuel and air must not be fully premixed as in true HCCI combustion. Both charge stratification and more premixed combustion can be obtained by injection of fuels in the gasoline autoignition range relatively late in the cycle compared to HCCI combustion. These issues will be discussed in detail in Chapter 6.

4.4 Should There Be a New Test to Rate the Autoignition Quality of Fuels?

A question that crops up often is whether the ASTM RON and MON tests need to be revised. There have been suggestions to change the operating conditions of the RON and MON tests [4.57] to make them more relevant to modern applications where the *K* value might be negative. However, this will not get over the problem that a practical fuel will have different ratings at different operating conditions. The main problem with the RON and MON tests is that the PRFs used to define the octane scale do not properly reflect the autoignition behavior of practical fuels. One possibility might be to use the ASTM RON test at higher engine pressures, but, more importantly, to use a toluene/n-heptane mixture (toluene reference fuel, or TRF) as the reference scale. In this scale, toluene would have an assigned value of 100, and n-heptane would be assigned a value of 0. A gasoline would be tested in the boosted CFR RON test and assigned a toluene number (TN), the volume percentage of toluene in the TRF that is closest to the gasoline for knock behavior in the revised RON test. Figure 4.19 shows the RON and MON of TRF plotted against volume percent of n-heptane.

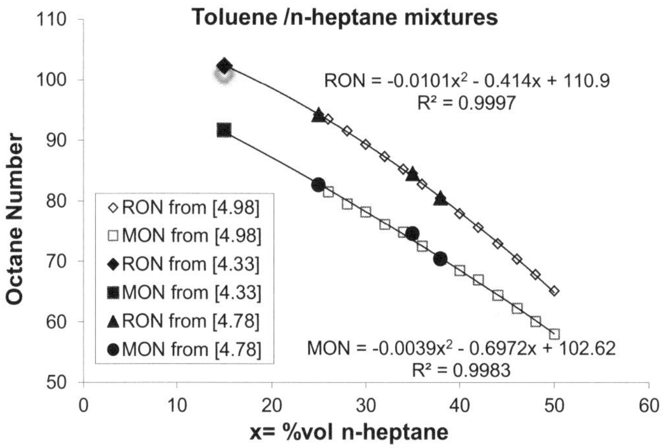

Fig. 4.19. RON and MON of mixtures of toluene and n-heptane, toluene reference fuels (TRFs).

Most of the data are from [4.98] and additional points with n-heptane concentrations of 15, 25, 35, and 38% by volume are from [4.33] and [4.78]. Second-order polynomial fits through all these data are also shown in Fig. 4.19. The measured sensitivities of TRFs lie in the range of 9.5 to 12 for RON > 75 (see Fig. 4.20), and hence would be very similar to the sensitivities of practical gasolines.

Since a TRF of the same RON as a gasoline would have almost the same MON, the *OI* for the two fuels would be comparable over a wide range of *K* values and the TN in the modified CFR RON test would describe the autoignition behavior of the gasoline very much better than its RON or MON through all operating conditions. To illustrate this point, consider a U.S. regular gasoline of 92 RON and 82 MON – (RON + MON)/2 of 87. A TRF with a TN of 72.5 (27.5% volume n-heptane) would have a RON and MON of 91.9 and 80.5, respectively (Fig. 4.19).

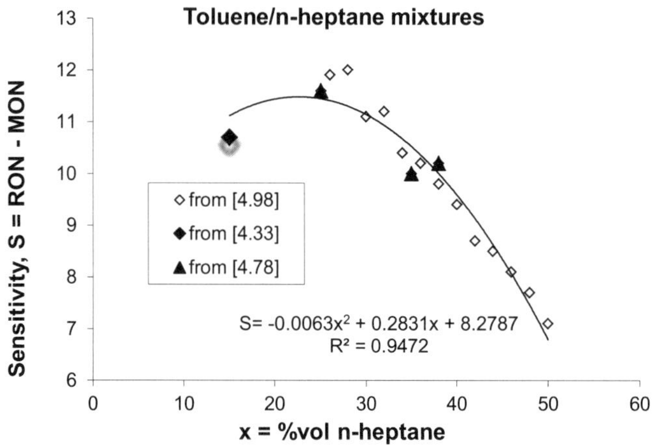

Fig. 4.20. Sensitivity of mixtures of toluene and n-heptane, toluene reference fuels (TRFs).

OI for these two fuels at different *K* values are listed in Table 4.2 and plotted in Fig. 4.21. It can be seen that the TRF with a TN of 72.5 has comparable OI to that of the actual gasoline over a wide range of *K* values, and so it can be expected to represent the autoignition behavior of the gasoline over a wide operating range.

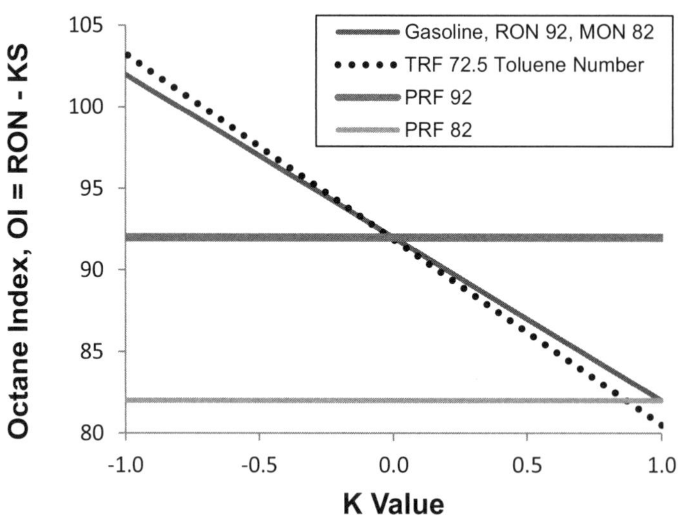

Fig. 4.21. *OI* versus *K* for a gasoline of 92 RON and 82 MON, TRF of 72.5 TN (RON/MON of 91.9/80.5), PRF 92, and PRF 82. The TRF tracks the OI of the gasoline across different operating conditions, that is, different K values, quite well.

Table 4.2 OI = RON − K(RON − MON), of a U.S. Regular Gasoline Compared to a TRF Fuel with a TN of 72.5

K value	OI of U.S. Regular Gasoline, (RON + MON)/2 of 87	OI of TRF with a TN of 72.5 (27.5% vol. n-heptane)
+1 (MON test Cond.)	82.0	80.5
+0.5	87.0	86.2
+0.2	90.0	89.6
0 (RON test Cond.)	92.0	91.9
-0.2	94.0	94.2
-0.5	97.0	97.6
-1.0	102.0	103.3

A single TRF rating, in this particular case of 72.5 TN, would be sufficient to characterize the gasoline for autoignition quality reasonably well at different operating conditions. In contrast, each of the PRFs matches the autoignition behavior of the gasoline (i.e., has the same *OI*), only at one operating condition (*K* value). Thus, PRF 92 matches the autoignition behavior of the gasoline only at the RON test condition ($K = 0$), while PRF 82 matches the gasoline only at the MON test condition ($K = 1$).

There might be several possible methods of ranking fuels in the new test:

1. Fix the intake pressure to, say, 1.5 bar (0.15 MPa). Run the CFR engine on different TRFs to establish the scaling between compression ratio at knock and TN with all other operating conditions fixed as in the current RON test. Run the gasoline to be rated at the same conditions, and find the compression ratio at knock and assign the TN from the scaling law.
2. Fix the compression ratio of the CFR engine to, say, 8.0 and the spark timing to, say, 10 CAD BTDC (before top dead center) with other conditions as in the RON test. These values will have to be chosen so that n-heptane at one end of the new scale will knock without any boost in the engine. Run different TRFs at different boost conditions to establish the scaling between TN and absolute intake pressure. Run the gasoline to be rated at the same condition, find the intake pressure at which knock is observed, and assign the TN from the scaling law.

Currently, high-octane fuels such as ethanol blends are difficult to rate quantitatively because their RON is much higher than 100 and so well outside the octane scale. A scale based on TRF would allow such fuels to be rated quantitatively with greater confidence. Running the modified RON test at a higher intake pressure would place the test in an operating regime that is more relevant to modern engines (i.e., beyond RON; see Section 4.2) and enables such knock-resistant fuels to be rated more easily. Indeed, any such new test could use a more modern single-cylinder engine, but the advantage of using the CFR engine is that it is in use across the world, and the ASTM RON test procedure would require only a slight modification. Though such a new test would be more appropriate than the RON or MON tests for practical gasolines, it will be less suited for paraffinic fuels in the gasoline autoignition range. Paraffinic fuels will have different TNs at different operating conditions. For instance, iso-octane will have a TN of 75—it will match a toluene reference fuel with 75% toluene (RON/MON = 94.2/82.8) when $K = -0.5$, but will most likely have a TN 89.3 (RON/MON =105.3/94.7) when $K = +0.5$. However, the RON and MON tests are embedded deeply into the oil, and to some extent in auto industries, and we should expect that they won't be replaced unless the arguments for doing so are compelling.

4.5 Summary and Concluding Remarks

Knock is the main constraint on the efficiency of SI engines, and antiknock quality is the primary quality parameter for fuels in SI engines. In both SI engines and HCCI engines, fuel and air are fully premixed, but the mixture is never truly homogeneous because of local temperature and mixture strength gradients. Knock in SI engines and combustion in HCCI engines start with autoignition at localized hot spots, and the evolution of the autoignition front through this inhomogeneity determines the amplitude of the resultant pressure wave. However, in most engine applications, the primary aim is to determine when, in the engine cycle, autoignition occurs. For this purpose, it is adequate to assume that the mixture is homogeneous.

Autoignition is determined by the chemical kinetics and the pressure and temperature history of the mixture. Fundamental studies such as measurements of ignition delay in shock tubes and rapid compression machines and modeling of chemical kinetics have helped greatly in understanding autoignition. Practical fuels are too complex for comprehensive chemical kinetic schemes to be developed for them. Chemical kinetic models are being developed for surrogate fuels, which are blends of a few pure compounds but match the behavior of a practical fuel. Given the central role of RON and MON in the fuel industry, regardless of any advances in fundamental understanding of autoignition and the accuracy and the applicability of kinetic models, autoignition in engines using practical fuels will need to be described and understood in terms of RON and MON of the fuel in the foreseeable future.

However, the RON and MON scale is based on PRFs whose chemistry is very different from that of other fuels. RON and MON do not describe the autoignition quality of practical fuels away from the RON or MON test conditions. In engines, the variation of the temperature in the unburned gas with pressure as described by T_{comp15} in Fig. 4.10 is of central importance. HCCI engines can be run over a very wide range of T_{comp15}. SI engines have moved progressively from near MON conditions toward and then beyond RON conditions as their efficiency and power density have improved through history. The true antiknock quality of a fuel is best described by *OI* (Eq. 4.8), which is a combination of RON and MON and a parameter related to engine design and operating conditions, *K*, which decreases as T_{comp15} decreases. For all practical purposes, if two fuels have the same *OI*, they have the same behavior in knocking SI engines or in HCCI engines. Therefore, a good surrogate fuel is the one that has the same RON and MON as the target gasoline; the two fuels will have the same *OI* whatever the value of *K*. A toluene, iso-octane, and n-heptane mixture can be found to match any RON and MON combination [4.99] and will be a good surrogate fuel for gasoline. In modern SI engines, *K* is negative, so that for a given RON, a fuel with lower MON has better antiknock quality. This is consistent with the fact that as pressure increases for a given temperature, nonparaffinic fuels become relatively more resistant to autoignition. Why this is so is not yet understood in terms of chemical kinetics. Everything that is done to improve the efficiency of SI engines will take them beyond RON, so that future engines will have negative *K* values. The implications of this will be discussed in Chapter 7.

4.6 References

4.1 Westbrook, C.K. 2000. "Chemical Kinetics of Hydrocarbon Ignition in Practical Combustion Systems." *Proceedings of the Combustion Institute* 28:1563–1577.

4.2 Kalghatgi, G.T., Risberg, P., and Ångström, H.-E. 2003. "A Method of Defining Ignition Quality of Fuels in HCCI Engines." SAE Paper No. 2003-01-1816. SAE International, Warrendale, PA.

4.3 Westbrook, C.K., and Dryer, F.L. 1984. "Chemical Kinetic Modelling of Hydrocarbon Combustion." *Progress in Energy and Combustion* Science 10:1.

4.4 Warnatz, J. 1984. "Chemistry of High Temperature Combustion of Alkanes Up to Octane." Paper presented at the Twentieth Symposium (International) on Combustion. The Combustion Institute, Pittsburgh, PA, p. 845.

4.5 Simmie, J.M. 2003. "Detailed Chemical Kinetic Models for the Combustion of Hydrocarbon Fuels." *Progress in Energy and Combustion Science* 29:599.

4.6 Bradley, D., and Morley, C. 1997. "Autoignition in Spark-Ignition Engines." In *Low-Temperature Combustion And Autoignition* (chap. 7), edited by M.J. Pilling, Vol. 35 of *Comprehensive Chemical Kinetics*, edited by R.G. Compton and G. Hancock, Elsevier Science B.V.

4.7 Griffiths, J.F. 1995. "Reduced Kinetic Models and Their Application to Practical Combustion Systems." *Progress in Energy and Combustion* Science 21:25.

4.8 Farrell, J.T., Cernansky, N.P., Dryer, F.L., Friend, D.G., Hergart, C.A., Law, C.K., McDavid, R., Mueller, C.J., Patel, A.K., and Pitsch, H. 2007. "Development of an Experimental Database and Kinetic Models for Surrogate Diesel Fuels." SAE Paper No. 2007-01-0201. SAE International, Warrendale, PA.

4.9 Pitz, W.J., Cernansky, N.P., Dryer, F.L., Egolfopoulos, F., Farrell, J.T., Friend, D.G., and Pitsch, H. 2007. "Development of an Experimental Database and Kinetic Models for Surrogate Gasoline Fuels." SAE Paper No. 2007-01-0175. SAE International, Warrendale, PA.

4.10 Naik, C.V., et al. 2005. "Detailed Chemical Kinetic Modeling of Surrogate Fuels for Gasoline and Application to an HCCI Engine." SAE Paper No. 2005-01-3741. SAE International, Warrendale, PA.

4.11 Pudupakkam, K.V., Naik, C.V., Wang, C., and Meeks, E. 2010. "Validation Studies of a Detailed Kinetics Mechanism for Diesel and Gasoline Surrogate Fuels." SAE Paper No. 2010-01-0545. SAE International, Warrendale, PA.

4.12 Andrae, J., Brinck, T., and Kalghatgi, G.T. 2008. "HCCI Experiments with Toluene Reference Fuels Modeled by a Semi-Detailed Chemical Kinetic Model." *Combustion and Flame* 155:696–712.

4.13 Andrae, J. 2008. "Development of a Detailed Chemical Kinetic Model for Gasoline Surrogate Fuels." *Fuel* 87:2013–2022.

4.14 Livengood, J.C., and Wu, P.C. 1955. "Correlation of Autoignition Phenomenon in Internal Combustion Engines and Rapid Compression Machines." In *Proceedings of Fifth International Symposium on Combustion*, Reinhold, p. 347.

4.15 Kalghatgi, G.T., Nakata, K., and Mogi, K. 2005. "Octane Appetite Studies in Direct Injection Spark Ignition (DISI) Engines." SAE Paper No. 2005-01-0244. SAE International, Warrendale, PA.

4.16 Bradley, D., Morley, C., and Walmsley, H. 2004. "Relevance of Research and Motor Octane Numbers to the Prediction of Engine Autoignition." SAE Paper No. 2004-01-1970. SAE International, Warrendale, PA.

4.17 Fieweger, K., Blumenthal, R., and Adomeit, G. 1997. "Self-Ignition of SI Engine Model Fuels: A Shock Tube Investigation at High Pressure." *Combustion and Flame* 109:599–619.

4.18 Gauthier, B.M., Davidson, D.F., and Hanson, R.K. 2004. "Shock Tube Determination of Ignition Delay Times in Full-Blend and Surrogate Fuel Mixtures." *Combustion and Flame* 139:300–311.

4.19 Herzler, J., Fikri, M., Hitzbleck, K., Starke, R., Schulz, C., Roth, P., and Kalghatgi, G.T. 2007. "Shock-Tube Study of the Ignition of N-Heptane/Toluene/Air Mixtures at Intermediate Temperatures and High Pressures." *Combustion and Flame* 149:25–31.

4.20 Fikri, M., Herzler, J., Starke, R., Schulz, C., Roth, P., and Kalghatgi, G.T. 2008. "Autoignition of Gasoline Surrogate Mixtures at Intermediate Temperatures and High Pressures." *Combustion and Flame* 152:276–281.

4.21 Eng, J.A., Leppard, W.R., Najt, P.A., and Dryer, F.L. 1997. "The Interaction Between Nitric Oxide and Hydrocarbon Chemistry in a Spark Ignition Engine." SAE Paper No. 972889. SAE International, Warrendale, PA.

4.22 Risberg, P., Johansson, D., Andrae, J., Kalghatgi, G., Björnbom, P., and Ångström, H.-E. 2006. "The Influence of NO on Combustion Phasing in a HCCI Engine." SAE Paper No. 2006-01-0416. SAE International, Warrendale, PA.

4.23 Burluka, A.A., Liu, K., Sheppard, C.G.W., Smallbone, A.J., and Wooley, R. 2004. "The Influence of Simulated Residual and NO Concentrations on Knock Onset for PRFs and Gasolines." SAE Paper No. 2004-01-2998. SAE International, Warrendale, PA.

4.24 Dagaut, P., and Nicolle, A., 2005. "Experimental Study and Detailed Kinetic Modelling of the Effect of Exhaust Gas on Fuel Combustion: Mutual Sensitisation of the Oxidation of Nitric Oxide and Methane Over Extended Temperature and Pressure Ranges." *Combustion and Flame* 140:161–171.

4.25 Leppard, W.R. 1990. "the Chemical Origin of Fuel Octane Sensitivity." SAE Paper No. 902137. SAE International, Warrendale, PA.

4.26 Kalghatgi, G.T., Golombok, M., and Snowdon, P. 1995. "Fuel Effects on Knock, Heat Release and 'CARS' Temperatures in a Spark-Ignition Engine." *Combustion Science and Technology* 110–111:209–228

4.27 Yates, A.D., Swarts, A., and Viljoen, C.L. 2005. "Correlating Auto-Ignition Delays and Knock-Limited Spark-Advance Data for Different Types of Fuel." SAE Paper No. 2005-01-2083. SAE International, Warrendale, PA.

4.28 Kalghatgi, G.T., and Head, R.A. 2004. "The Available and Required Autoignition Quality of Gasoline-Like Fuels in HCCI Engines at High Temperatures." SAE Paper No. 2004-01-1969. SAE International, Warrendale, PA.

4.29 Kalghatgi, G.T., and Head, R.A. 2006. "Combustion Limits and Efficiency in an HCCI Engine." *International Journal of Engine Research* 7(3):215–236.

4.30 Heywood, J.B. 1988. Chap. 9. In *Internal Combustion Engine Fundamentals.* McGraw-Hill, New York.

4.31 Young, M.B. 1981. "Cyclic Dispersion in the Homogeneous-Charge Spark-Ignition Engine-A Literature Survey." SAE Paper No. 810020. SAE International, Warrendale, PA.

4.32 Kalghatgi, G.T. 1987. "Spark Ignition, Early Flame Development and Cyclic Variation in I.C. Engines." SAE Paper No. 870163. SAE International, Warrendale, PA.

4.33 Kalghatgi, G.T., Snowdon, P., and McDonald, C.R. 1995. "Studies of Knock in a Spark Ignition Engine with CARS Temperature Measurements and Using Different Fuels." SAE Paper No. 950690. SAE International, Warrendale, PA.

4.34 Burgdorf, K., and Denbratt, I. 1997. "Comparison of Cylinder Pressure Based Knock Detection Methods." SAE Paper No. 972932. SAE International, Warrendale, PA.

4.35 Gao, X., Stone, R., Hudson, C., and Bradbury, I. 1993. "The Detection and Quantification of Knock in Spark Ignition Engines." SAE Paper No. 932759. SAE International, Warrendale, PA.

4.36 Collings, N., Dinsdale, S., Eade, D., 1986. "Knock Detection by Means of the Spark Plug." SAE Paper No. 860635. SAE International, Warrendale, PA.

4.37 Nakagawa, Y., Takagi, Y., Itoh, T., and Iijima, T. 1984. "Laser Shadowgraphic Analysis of Knocking in S.I. Engine." SAE Paper No. 845001. SAE International, Warrendale, PA.

4.38 König, G., and Sheppard, C.G.W. 1990. "End Gas Autoignition and Knock in a Spark Ignition Engine." SAE Paper No. 902135. SAE International, Warrendale, PA.

4.39 Kalghatgi, G.T. 2001. "Fuel Anti-knock Quality—Part I. Engine studies." SAE Paper No. 2001-01-3584. SAE International, Warrendale, PA.

4.40 Franklin, M.L., and Murphy, T.E. 1989. "A Study of Knock and Power Loss in the Automotive Spark Ignition Engine." SAE Paper No. 890161. SAE International, Warrendale, PA.

4.41 Chun, K.M., and Heywood, J.B. 1989. "Characterization of Knock in a Spark-Ignition Engine." SAE Paper No. 890156. SAE International, Warrendale, PA.

4.42 Bradley, D., Kalghatgi, G.T., and Golombok, M. 1996. "Fuel Blend and Mixture Strength Effects on Autoignition Heat Release Rates and Knock Intensity in S.I. Engines." SAE Paper No. 962105. SAE International, Warrendale, PA.

4.43 Bradley, D., and Kalghatgi, G.T. 2009. "Influence of Autoignition Delay Time Characteristics of Different Fuels on Pressure Waves and Knock in Reciprocating Engines." *Combustion and Flame* 156:2307–2318.

4.44 König, G., Maly, R.R., Bradley, D., Lau, A.K.C. and Sheppard, C.G.W. 1990. "Role of Exothermic Centres on Knock Initiation and Knock Damage." SAE Paper No. 902136. SAE International: Warrendale, PA.

4.45 Bradley, D. 2012. "Autoignitions and Detonations in Engines and Ducts." *Philosophical Transactions A of the Royal Society A* 370:689–714.

4.46 Kalghatgi, G.T. 2001. "Fuel anti-knock—Part II. Vehicle Studies—How Relevant Is Motor Octane Number (MON) for Modern Engines?" SAE Paper No. 2001-01-3585. SAE International, Warrendale, PA.

4.47 Bell, A. 2010. "Modern SI Engine Control Parameter Responses and Altitude Effects with Fuels of Varying Octane Sensitivity." SAE Paper No. 2010-01-1454. SAE International, Warrendale, PA.

4.48 McNally, M.J., Callison, J.C., Evans, B., Graham, J.P., Swaynos, D.V., Uihlein, J.P., and Wusz, T. 1991. "The Effect of Gasoline Octane Quality on Vehicle Acceleration Performance—A CRC Study." SAE Paper No. 912304. SAE International, Warrendale, PA.

4.49 Coordinating Research Council. 1995. "Vehicle Performance Effect of Octane on Knock Sensor Equipped Vehicles." CRC Report No. 597.

4.50 Whelan, D.E., Schmidt, G.K., and Hischier, M.E. 1998. "An Acceleration Based Method to Determine the Octane Number Requirement of Knock Sensor Vehicles." SAE Paper No. 982721. SAE International, Warrendale, PA.

4.51 Sugawara, Y., Akasaka, Y., and Kagami, M. 1997. "Effects of Gasoline Properties on Acceleration Performance of Commercial Vehicles." SAE Paper No. 971725. SAE International, Warrendale, PA.

4.52 Bell, A.G. 1975. "The Relationship Between Octane Quality and Octane Requirement." SAE Paper No. 750935. SAE International, Warrendale, PA.

4.53 Arrigoni, V., Cornetti, G.M., Spallanzani, G., Calvi, F., and Tontodonati, A. 1974. "High Speed Knock in S.I. Engines." SAE Paper No. 741056. SAE International, Warrendale, PA.

4.54 Ingamells, J.C., and Jones, E.R. 1981. "Developing Road Octane Correlations from Octane Requirement Surveys." SAE Paper No. 810492. SAE International, Warrendale, PA.

4.55 Douaud, A., and Eyzat, P. 1978. "Four-Octane-Number Method for Predicting the Anti-Knock Behavior of Fuels and Engines." SAE Paper No. 780080. SAE International, Warrendale, PA.

4.56 Kalghatgi, G.T. 2005. "Auto-Ignition Quality of Practical Fuels and Implications for Fuel Requirements of Future SI and HCCI Engines." SAE Paper No. 2005-01-0239. SAE International, Warrendale, PA.

4.57 Mittal, V., and Heywood, J.B. 2008. "The Relevance of Fuel RON and MON to Knock Onset in Modern SI Engines." SAE Paper No. 2008-01-2414. SAE International, Warrendale, PA.

4.58 CRC. 2011. "Fuel Antiknock Quality—Engine Response to RON vs MON, Scoping Tests—Final Report." Report No. 660. Coordinating Research Council (CRC).

4.59 Gerty, M.D., and Heywood, J.B. 2006. "An Investigation of Gasoline Engine Knock Limited Performance and the Effects of Hydrogen Enhancement." SAE Paper No. 2006-01-0228. SAE International, Warrendale, PA.

4.60 Akihama, K., Taki, M., Takasu, S., Ueda, T., Iwashita, Y., Farrell, J.T., and Weissman, W. 2004. "Fuel Octane and Composition Effects on Efficiency and Emissions in a High Compression Ratio SIDI Engine." SAE Paper No. 2004-01-1950. SAE International, Warrendale, PA.

4.61 Amer, A., Babiker, H., Chang, J., Kalghatgi, G., Adomeit, P., Brassat, A., and Guenther, M., 2012. "Fuel Effects on Knock in a Highly Boosted Direct Injection Spark Ignition Engine." SAE Paper No. 2012-01-1634. SAE International, Warrendale, PA.

4.62 Davies, T., Cracknell, R., Lovett, G., Cruff, L., and Fowler, J. 2011. "Fuel Effects in a Boosted DISI Engine." SAE Paper No. 2011-01-1985. Also JSAE 20119155. SAE International, Warrendale, PA.

4.63 Mittal, V., and Heywood, J.B. 2009. "The Shift in Relevance of Fuel RON and MON to Knock Onset in Modern SI Engines Over the Last 70 Years." SAE Paper No. 2009-01-2622. SAE International, Warrendale, PA.

4.64 Najt, P., and Foster, D. 1983. "Compression-Ignited Homogeneous Charge Combustion." SAE Technical Paper No. 830264. SAE International, Warrendale, PA.

4.65 Thring, R. 1989. "Homogeneous-Charge Compression-Ignition (HCCI) Engines." SAE Paper No. 892068. SAE International, Warrendale, PA.

4.66 Christensen, M., Johansson, B., and Einewall, P. 1997. "Homogeneous Charge Compression Ignition (HCCI) Using Isooctane, Ethanol and Natural Gas—A Comparison with Spark Ignition Operation." SAE Paper No. 972874. SAE International, Warrendale, PA.

4.67 Stanglmaier, R., and Roberts, C. 1999. "Homogeneous Charge Compression Ignition (HCCI): Benefits, Compromises, and Future Engine Applications." SAE Paper No. 1999-01-3682. SAE International, Warrendale, PA.

4.68 Christensen, M., Hultqvist, A., and Johansson, B. 1999. "Demonstrating the Multi Fuel Capability of a Homogeneous Charge Compression Ignition Engine with Variable Compression Ratio." SAE Paper No. 1999-01-3679. SAE International, Warrendale, PA.

4.69 Zhao, H. (editor). 2007. *HCCI and CAI Engines*. Woodhead, Cambridge, UK.

4.70 Hultqvist, A., Christensen, M., Johansson, B., Richter, M., et al. 2002. "The HCCI Combustion Process in a Single Cycle—Speed Fuel Tracer LIF and Chemiluminescence Imaging." SAE Paper No. 2002-01-0424. SAE International, Warrendale, PA.

4.71 Reuss, D., and Sick, V. 2005. "Inhomogeneities in HCCI Combustion: An Imaging Study." SAE Paper No. 2005-01-2122. SAE International, Warrendale, PA.

4.72 Bradley, D., Morley, C., Gu, X.J., and Emerson, D.R. 2002. "Amplified Pressure Waves During Auto-Ignition: Relevance to CAI Engines." SAE Paper No. 2002-01-2868. SAE International, Warrendale, PA.

4.73 Sheppard, C.G.W, Tolegano, S., and Wooley, R. 2002. "On the Nature of Auto-Ignition Leading to Knock in HCCI Engines." SAE Paper No. 2002-01-2831. SAE International, Warrendale, PA.

4.74 Oakley, A., Zhao, H., Ladommatos, N., and Ma, T. 2001. "Dilution Effects on the Controlled Auto-Ignition (CAI) Combustion of Hydrocarbon and Alcohol Fuels." SAE Paper No. 2001-01-3606. SAE International, Warrendale, PA.

4.75 Montagne, X., and Duret, P. 2002. "What Will Be the Future Combustion and Fuel-Related Technology Challenges?" In *A New Generation of Engine Combustion Processes for the Future?* edited by P. Duret, Editions Technip, Paris, p 177.

4.76 Aroonsrisopon, T., Sohm, V., Werner, P., Foster, D. et al. 2002. "An Investigation into the Effect of Fuel Composition on HCCI Combustion

Characteristics." SAE Paper No. 2002-01-2830. SAE International, Warrendale, PA.

4.77 Ryan, T.W., and Matheaus, A. 2002. "Fuel Requirements for HCCI Engine Operation." *Proceedings of THIESEL 2002, Thermo and Fluid Dynamic Processes in Diesel Engines, Valencia.*

4.78 Risberg, P., Kalghatgi, G.T., and Ångström H.-E. 2003. "Autoignition Quality of Gasoline-Like Fuels in HCCI Engines." SAE Paper No. 2003-01-3215. SAE International, Warrendale, PA.

4.79 Jeuland, N., Montagne, X., and Duret, P. 2003. "Engine and Fuel Related Issues of Gasoline CAI (Controlled Auto-Ignition) Combustion." SAE Paper No. 2003-01-1856. SAE International, Warrendale, PA.

4.80 Koopmans, L., Strömberg, E., and Denbratt, I. 2004. "The Influence of PRF and Commercial Fuels with High Octane Number on the Auto-ignition Timing of an Engine Operated in HCCI Combustion Mode with Negative Valve Overlap." SAE Paper No. 2004-01-1967. SAE International, Warrendale, PA.

4.81 Risberg, P., Kalghatgi, G.T., and Ångström H.-E. 2004. "The Influence of EGR on Autoignition Quality of Gasoline-Like Fuels in HCCI Engines." SAE Paper No. 2004-01-2952. SAE International, Warrendale, PA.

4.82 Shibata, G., and Urushihara, T. 2006. "The Interaction Between Fuel Chemicals and HCCI Combustion Characteristics Under Heated Intake Air Conditions." SAE Paper No. 2006-01-0207. SAE International, Warrendale, PA.

4.83 Sjöberg, M., and Dec, J. 2003. "Combined Effects of Fuel-Type and Engine Speed on Intake Temperature Requirements and Completeness of Bulk-Gas Reactions for HCCI Combustion." SAE Paper No. 2003-01-3173. SAE International, Warrendale, PA.

4.84 Kalghatgi, G.T. 2007. "Fuel Effects in Gasoline HCCI Engines." In *HCCI and CAI Engines*, edited by H. Zhao, Woodhead, Cambridge, UK, pp. 206–238.

4.85 Bradley, D., and Head, R.A. 2006. "Engine Autoignition: The Relationship Between Octane Numbers and Autoignition Delay Times." *Combustion and Flame* 147:171–184.

4.86 Eng, J. 2002. "Characterization of Pressure Waves in HCCI Combustion." SAE Paper No. 2002-01-2859. SAE International, Warrendale, PA.

4.87 Shibata, G., Oyama, K., Urushihara, T., and Nakano, T. 2004. "The Effect of Fuel Properties on Low and High Temperature Heat Release and Resulting Performance of an HCCI Engine." SAE Paper No. 2004-01-0553. SAE International, Warrendale, PA.

4.88 Shibata, G., Oyama, K., Urushihara, T., and Nakano, T. 2005. "Correlation of Low Temperature Heat Release With Fuel Composition and HCCI Engine Combustion." SAE Paper No. 2005-01-0138. SAE International, Warrendale, PA.

4.89 Shibata, G., and Urushihara, T. 2007. "Auto-Ignition Characteristics of Hydrocarbons and Development of HCCI Fuel Index." SAE Paper No. 2007-01-0220. SAE International, Warrendale, PA.

4.90 Ryan, T.W. 2007. "HCCI Fuel Requirements." In *HCCI and CAI Engines*, edited by H. Zhao, Woodhead, Cambridge, UK, pp. 342–362.

4.91 Ryan, T., and Matheaus, A. 2003. "Fuel Requirements for HCCI Engine Operation." SAE Paper No. 2003-01-1813. SAE International, Warrendale, PA.

4.92 Risberg, P., Kalghatgi, G.T., and Ångström H.-E. 2005. "Autoignition Quality of Diesel-Like Fuels in HCCI Engines." SAE Paper No. 2005-01-2127. SAE International, Warrendale, PA.

4.93 Griffiths, J.F., Halford-Maw, P.A., and Mohammed, C. 1997. "Spontaneous Ignition Delays as a Diagnostic of the Propensity of Alkanes to Cause Engine Knock." *Combustion and Flame* 111:327–337.

4.94 Tanaka, S., Ayala, F., Keck, J.C., and Heywood, J.B. 2003. "Two-Stage Ignition in HCCI Combustion and HCCI Combustion Control by Fuels and Additives." *Combustion and Flame* 132:219–239.

4.95 Aroonsrisopon, T., Werner, P., Waldman, J., Sohm, V., et al. 2004. "Expanding the HCCI Operation with the Charge Stratification." SAE Paper No. 2004-01-1756. SAE International, Warrendale, PA.

4.96 Berntsson, A., and Denbratt, I. 2007. "HCCI Combustion Using Charge Stratification for Combustion Control." SAE Paper No. 2007-01-0210. SAE International, Warrendale, PA.

4.97 Sjöberg, M., and Dec, J. 2005. "Effects of Engine Speed, Fueling Rate, and Combustion Phasing on the Thermal Stratification Required to Limit HCCI Knocking Intensity." SAE Paper No. 2005-01-2125. SAE International, Warrendale, PA.

4.98 *Annual Book of ASTM Standards*, Vol. 5.01–5.05. 2011. ASTM International, West Conshohocken, PA.

4.99 Morgan, N., Smallbone, A., Bhave, A., Kraft, M., Cracknell, R., and Kalghatgi, G. 2010. "Mapping Surrogate Gasoline Compositions into RON/MON Space." *Combustion and Flame* 157:1122–1113.

Chapter 5
Preignition and Super-Knock in Turbocharged Spark Ignition Engines

Turbocharged direct injection spark ignition (DISI) engines have been observed to suffer from an abnormal combustion phenomenon known as pre-ignition [5.1–5.9], in which an expanding flame front is established before the spark plug fires in the engine. Preignition has been of intermittent concern at different stages of the history of spark ignition (SI) engines [5.10–5.19]. In the 1950s and early 1960s, compression ratios increased rapidly, and preignition became a serious problem [e.g., 5.11–5.15]. In some quarters it was thought that preignition might become more important than knock as an obstacle to improved engine performance [5.14]. Preignition seems to have attracted attention again in the 1980s [e.g., 5.17–5.19] after a relatively quiet period in the 1960s and 1970s, primarily because of the problems encountered with using methanol. We first discuss preignition, which is the real problem, because without it, super-knock is not very likely to occur.

Preignition causes the cylinder pressure to increase above the compression pressure before the spark plug fires. The early start to combustion from pre-ignition causes the pressure and the temperature of the unburned gas ahead of the advancing flame front, the end-gas, to rise more rapidly compared to normal spark timing. If autoignition occurs in the end-gas at high pressure and temperature, it can lead to extremely heavy knock, sometimes described as "super-knock," another abnormal combustion phenomenon that could potentially damage the engine [5.1–5.8]. Note that there is no quantitative definition of super-knock. Examples of in-cylinder pressure from cycles, originally discussed in [5.7], showing preignition and super-knock, are shown in Fig. 5.1.

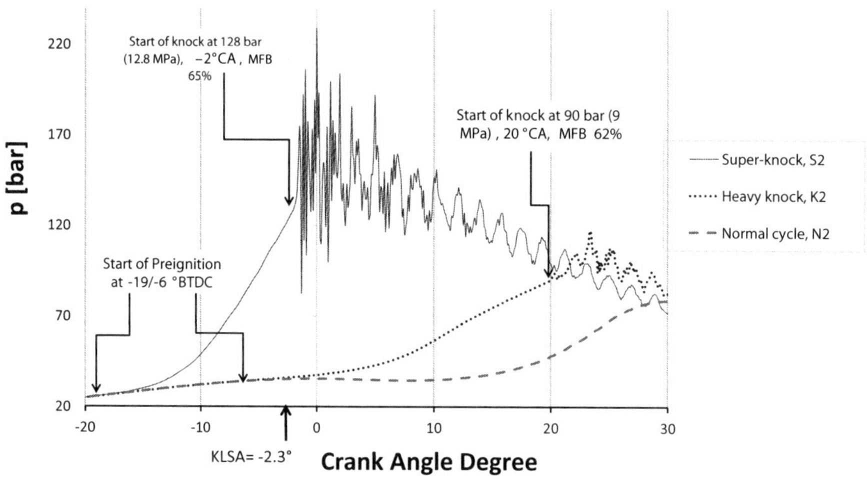

Fig. 5.1. Preignition and super-knock in a turbocharged engine. Examples of in-cylinder pressure traces are from experiments discussed in [5.7].

Cycle S2 shows knock intensity (KI), peak-to-peak pressure fluctuation, of over 140 bar (14 MPa) and could be considered as a super-knock event. In contrast, the "normal" knocking cycle, Cycle N2, which does not show preignition might have a KI of less than 0.5 bar (0.05 MPa). Cycle K2 has preignition starting later than S2 and a KI of about 24 bar (2.4 MPa), though at the time of the start of autoignition, it had a similar amount, about 40% of unburned mass fraction as Cycle S2. To put these knock intensities into context, in engine knock experiments it is customary to set the knock threshold at a KI of 0.2 to 0.5 bar (0.02 to 0.05 MPa), and a KI of 2 bar (0.2 MPa) would be considered heavy knock that is not knowingly exceeded for extended periods. An example of a super-knock event from a different engine also has been presented and discussed in [5.7].

Figure 5.2 shows the heat release rates estimated from the pressure for these three cycles. Such a calculation cannot be done for Cycles S2 and K2 once large pressure fluctuations set in.

The rate of heat release in Cycles S2 and K2 after the initiation of preignition, but before the knock event later in the cycle, is lower than the peak heat release rate in Cycle N2, which is a normal cycle without preignition. Heat release rates following preignition are comparable to those in a normal cycle with spark ignition at the selected spark timing and are much smaller than can

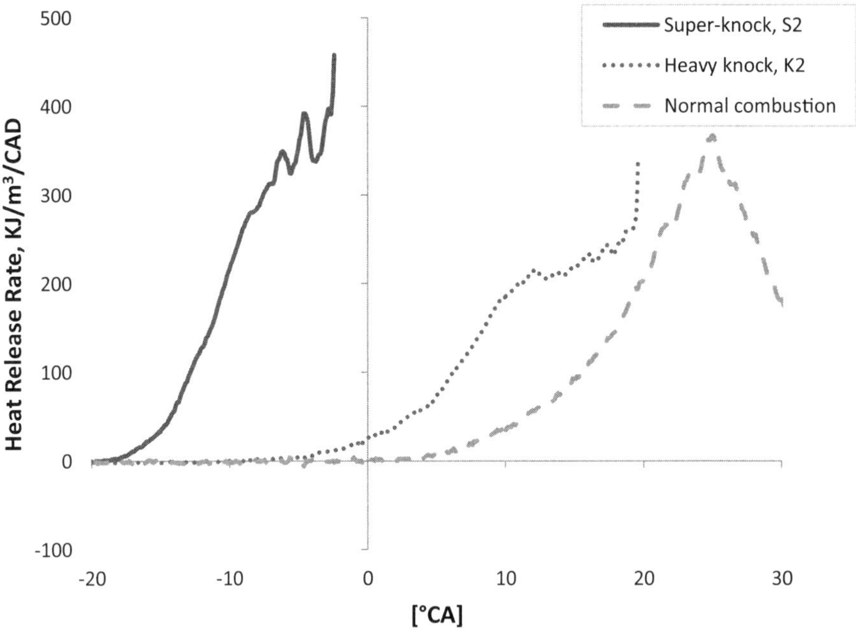

Fig. 5.2. Heat release rates calculated from the pressure signals in Fig. 5.1.

be expected from autoignition (Fig. 5.2; see also [5.2]). This suggests that heat release following the initiation of preignition occurs in an expanding flame front as in a normal cycle. Optical studies show the development of a turbulent flame following preignition (e.g., Fig. 6 in [5.4]; Fig. 3 in [5.9]). Ion gaps have confirmed flame development following preignition in engines without optical access [5.11, 5.13]. Thus, experimental evidence shows that a turbulent flame front is established after preignition.

These events are sporadic and infrequent—they might occur once in thousands of cycles [5.1–5.4]. The increased pressure during the compression stroke caused by preignition also leads to excessive negative work before the end of the compression stroke and results in loss of power and an increase in the rate of heat transfer to the combustion walls. In many circumstances, this increased heat transfer could intensify the preignition process [5.10] depending on how it originates. In a single-cylinder engine, this could eventually cause the engine to stop. In a multicylinder engine, with one cylinder suffering from preignition, the cylinders operating normally could keep the engine running. However, the high pressure and temperature developed in the compression stroke of the cylinder suffering from preignition could cause the failure of the piston or the connecting rod [5.10].

5.1 Preignition

The successful establishment of a flame front by an external energy source depends on the competition between chemical heat release and the loss of heat and radicals to the surroundings through conduction, diffusion, and radiation [5.20]. In the first instance, the fuel/air mixture in a localized, small volume needs to be heated to a sufficiently high temperature to cause a thermal runaway. Chemical reactions start consuming the reactant when this ignition criterion is fulfilled and a combustion wave starts to propagate outward [5.21].

To continue to propagate, the flame should have grown to a critical size by the time the temperature at the origin has decreased to around the normal flame temperature. Below this critical flame size, the rate of heat release within the inner zone of the chemical reaction is insufficient to compensate for the rate of heat loss to the surroundings, and the temperature would decrease throughout the reaction volume in the absence of an external energy input of sufficient power and duration. In such a case, the reaction would gradually cease and the flame would become extinct only after a small amount of fuel/air mixture is burned. To drive the flame kernel beyond the critical size, the energy input from the external source must satisfy a stronger initiation criterion.

Hence, for preignition to occur, both the ignition and initiation criteria have to be met before the spark plug fires.

5.1.1 Fulfillment of the Ignition Criterion

In a practical SI engine, the spark timing would be set so that autoignition of the fuel/air mixture—even in the end-gas ahead of the flame front where pressure and temperature are higher than during the compression stroke—is avoided, and therefore, knock is avoided. Hence, the initial heat release setting off the preignition cannot be caused by the local autoignition of the fuel/air mixture, because the ignition delay of the bulk mixture is too low at these conditions [5.2, 5.20].

5.1.1.1 Surface Ignition

In earlier studies preignition occurred in either carburetor or port fuel injector engines with relatively low compression ratios and no turbocharging. In all these cases, preignition was known to start at a hot spot on the internal surface of the engine such as an overheated valve or a spark plug [5.10]. In general, internal surfaces of all used engines are always covered by deposits of varying quantities and properties, depending on the fuel, lubricant, and engine operating conditions (see Chapter 3). In practice, combustion chamber deposits (CCDs) will be implicated in any surface ignition, though surface ignition can always be triggered in the absence of CCDs [5.11] (e.g., by using a heated wire; see [5.10]). Solid particles in the combustion chamber can also contribute to preignition. Melby et al. described experiments [5.13] where preignition was greatly enhanced when particles of deposits scraped from engines and mixtures of deposit components prepared in the laboratory were introduced into an engine. The increased heat transfer to the walls caused by preignition could further increase the temperature of the surface hot spot and could intensify the preignition process [5.10].

5.1.1.2 Ignition Away from Surfaces: Autoignitive Preignition

In contrast to earlier studies, modern engines in which preignition has again become a concern have direct injection of fuel and significantly higher in-cylinder pressures because of turbocharging. Empirical evidence shows that in these modern engines, preignition is not usually initiated at the internal surface of the combustion chamber [5.1–5.4, 5.9]. Optical investigations [5.1, 5.4, 5.9] show that preignition starts at different points distributed randomly across the cylinder away from the internal surfaces. Zahdeh et al. [5.4] found that preignition could be greatly reduced by avoiding fuel impingement on the wall. Dahnz et al. [5.1] discussed the many possible mechanisms that could lead to a local hot spot, which could then lead to preignition.

Based on such observations, the following mechanism has been suggested for the start of preignition [5.1, 5.2, 5.4]. If fuel droplets end up on the cylinder wall, they mix with the oil film and change its surface tension and viscosity. This oil/fuel mixture collects in the crevice volume of the top land of the piston. Droplets of this oil/fuel mixture will be more likely to be blown into the cylinder by the in-cylinder airflow because of the reduced surface tension. The oil/fuel mixture will also have lower resistance to autoignition than the fuel itself because the oil has very low resistance to autoignition—it resembles diesel fuel rather than gasoline. During the compression stroke, as the temperature and pressure increase in the cylinder, these droplets can vaporize, mix with oxygen, undergo autoignition reactions, and eventually burn to act as ignition centers.

Two conditions have to be met in order for ignition to occur at a hot spot [5.22]:

1. The ignition delay τ_i must be small enough for autoignition to occur at a hot spot on a sufficiently short time scale.
2. The size of the hot spot must exceed the critical radius for autoignition r_c.

Based on the ignition delays of a typical gasoline at the pressures and temperatures at which preignitions have been observed, it can be shown that autoignition of the bulk fuel/air mixture in the gas phase is impossible [5.1, 5.2, 5.22]. Thus, the initial hot spot must be formed around droplets of a mixture of oil and fuel. Even if it is assumed that this mixture in the droplet is as reactive as n-heptane, the ignition delays would not be low enough to lead to autoignition in the gas phase after the droplet has mixed with air [5.22]. It is possible that the oil is much more reactive than n-heptane.

It is also likely that catalytic reactions—not properly understood—also play a role in promoting local heat release centered round oil/fuel droplets. There is experimental evidence for the catalytic effect of metallic additives in the oil in promoting preignition. For instance, there was a significant reduction of preignition when a standard oil, containing normal levels of metallic additives, was replaced by an oil made with a synthetic base oil and low levels of metallic additives [5.4]. Metallic additives in the oil can increase the metallic content of deposits and promote preignition even when preignition is initiated by surface ignition as in early studies. Guibet and Duval [5.16] reported that metallic lubricant additives, particularly calcium and barium additives, significantly increase the probability of preignition. Swain et al. [5.19] found that in a methanol-fueled engine, a calcium-based oil additive increased the probability of surface ignition. Menrad et al. [5.17] observed that a coating of noble metals on spark plug electrodes promoted preignition in methanol-fueled engines.

The detailed mechanisms by which such local heat release, centered round oil/fuel droplets, occurs are not known but are likely to involve both chemical kinetic and catalytic reactions. These chemical reaction rates are likely to be higher for higher pressures, so that the likelihood of local heat release occurring increases as the in-cylinder pressure is increased because of turbocharging.

A typical automotive coil-based ignition system delivers around 100 millijoules (mJ) of electrical energy through the spark, of which around 90% is lost to the electrodes and is not available to the mixture to initiate combustion [5.23]. Liquid hydrocarbons used in fuels and lubricants typically have a lower heating value of around 32 MJ/L. Hence, a spherical droplet of fuel and oil of a diameter of 100 microns (0.1 mm), would release around 17 mJ of energy if it were to be completely burned. Thus such a droplet can provide more energy than a typical spark and plausibly set off a flame if all other conditions are satisfied.

5.1.2 Fulfillment of the Initiation Criterion

The analysis of Zeldovich et al. [5.24] introduces the concept of a "flame ball" with a critical radius R_f. A spherical flame with a radius smaller than R_f extinguishes, while a spherical flame with a radius larger than R_f continues to propagate. Further, R_f is related to the laminar flame thickness δ by [5.24, 5.25],

$$\frac{R_f}{\delta} = \exp\frac{1}{2}\beta\left(1-\frac{1}{Le}\right) \tag{5.1}$$

Here, Le is the Lewis number, and β is the Zeldovich number = $E(T_b - T_u)/RT_b^2$. R_f is greater than δ if $Le > 1$ and smaller than δ if $Le < 1$. In any case, the critical radius R_f depends on δ. The larger the R_f, and from Eq. 5.1, the larger the δ, the more difficult it is to establish a propagating flame. Therefore, we should expect that the probability of flame initiation, once the ignition criterion is met, will decrease as δ increases.

The laminar flame thickness δ is given by $\delta = D/S_L$, where D is the thermal diffusivity and S_L is the laminar burning velocity. Further, $D = \mu/\rho$, where μ is the dynamic viscosity (assuming Prandtl number = 1). Hence,

$$\delta = (\mu/\rho S_L) \tag{5.2}$$

All else being equal, the larger the value of S_L, the higher is the probability of preignition.

Further [5.26–5.28], the dependence on pressure and temperature of S_L can be expressed in general as:

$$S_L = S_{L0}(T/T_0)^n (P/P_0)^{-m} \tag{5.3}$$

where S_{L0} is the laminar burning velocity at some reference pressure P_0 and temperature T_0.

Also, from ideal gas law, $(P/P_0) = (T/T_0)(\rho/\rho_0)$ where ρ_0 is the density at pressure P_0 and temperature T_0. Finally, dynamic viscosity μ varies as $T^{0.5}$.

If μ_0 is the dynamic viscosity at P_0 and T_0, from Eqs. 5.2 and 5.3,

$$\delta = (\mu_0/\rho_0 S_{L0})(P/P_0)^{(m-1)}(T/T_0)^{(1.5-n)} \tag{5.4}$$

Different values have been reported for different fuels in the literature for m and n. Metghalchi and Keck [5.26] suggested, for stoichiometric mixtures of iso-octane, n of 2.18 and m of 0.16. Gulder [5.27] suggested n of 1.75 for methanol and ethanol and 1.56 for iso-octane, and m varying between 0.17 and 0.22 for the three fuels. However, Bradley et al. gave a substantially lower value of 1.01 for n and a value of 0.282 for m, for iso-octane/air mixtures [5.28]. Bradley et al. suggest that the results in [5.26] and [5.27] are influenced by the increasing cellularity of the flame as the pressure is increased. In an engine, the conditions are more likely to be as in the experiments of Bradley et al. in [5.28]. Of course, S_{L0} also depends on stoichiometry [5.26, 5.27] and dilution. (e.g., S_{L0} decreases with increasing exhaust gas recirculation, EGR [5.26]). The values for m and n and the consequences for preignition are discussed in [5.22] for ethanol, hydrogen, iso-octane, and gasoline.

5.1.3 Methods Used for Quantifying Preignition

Cycles with preignition can be identified from the higher than normal pressure during the compression stroke or higher than normal maximum pressure [5.3]. Abnormal combustion patterns can also be recognized by using ion gaps in the combustion chamber to track flame development [5.11, 5.13]. The changes in the ion current in the spark gap have also been used to identify preignition [5.11, 5.18, 5.19]. The number of cycles with preignitions can be counted without in any way trying to promote preignition [5.1, 5.3, 5.4].

In some studies, preignition is promoted either by using spark plugs with a low heat range (e.g., with extended electrodes [5.14]) or by introducing a chemically inert pellet that is heated by the combustion gases [5.15]. A variation on this theme is to use the so-called postignition technique [5.17, 5.18]. In this method, the spark is disabled for several cycles, and preignition is triggered by the hot spark electrodes at a crank angle usually after the normal spark timing. Another method often used, particularly for rating fuels, is to introduce an electrically heated hot spot or glow plug in the engine [5.10, 5.11, 5.29, 5.30]. The method of assessing preignition resistance (PR) is described in [5.10]. The compression ratio and other operating conditions are fixed in a single-cylinder E6 engine. Preignition is initiated by an electrically heated hot wire in the combustion chamber. Fuels are rated according to the amount of electrical energy that has to be passed through the hot wire to cause preignition—the higher this energy, the higher the PR. A scale of quality according to PR is constructed where iso-octane is assigned the value 100 and cyclohexane is assigned 0. The fuels are tested at a mixture strength that is about 10% more fuel rich than stoichiometric—where laminar burning velocity can, in fact, be expected to be near its maximum for hydrocarbon fuels.

5.1.4 Pressure and Temperature Effects on Preignition

In turbocharged DISI engines, the pressure is higher than in non-turbocharged engines; engine designers attempt to keep the temperature similar to non-turbocharged engines. The autoignition reaction rates (and thus the chance of local heat release around oil/fuel droplets) and the probability meeting the ignition criterion can be expected to increase because of the increased pressure. If the intake temperature is fixed and the pressure increased, the temperature-dependent term in Eq. 5.4 will remain the same at a given crank angle during the compression stroke for a given fuel, but the pressure-related term will decrease because the pressure exponent in Eq. 5.4, $(m - 1)$, is negative. Hence, at a given crank angle, laminar flame thickness decreases and flame initiation becomes more likely as intake pressure increases, and therefore preignition becomes more likely in turbocharged engines for a given fuel, with all else being equal.

Clearly, one way to reduce the chance of preignition is to reduce the chance of oil/fuel droplets coming into the cylinder (e.g., by reducing the fuel impingement on the walls or by increasing the crevice volume in the top land of the piston [5.4]), if the combustion of the oil/fuel droplet is the dominant mechanism for reaching the ignition criterion. Similarly, removing metallic additives from the oil reduces the chances of preignition [5.4], most likely by reducing the probability of oil/fuel droplets igniting in the first place.

It has been found that increasing the temperature does not increase the probability of preignition. Experiments by Downs and Theobald [5.29] showed that significant changes in inlet air temperature and coolant temperature had little effect on preignition. In the experiments by Melby et al. [5.13], changing the inlet air temperature from 71 to 149 °C (160 to 300 °F) had little effect on preignition frequency. Menrad et al. [5.17] also found that increasing the intake air temperature did not have a significant effect on preignition. On the other hand, in [5.1] and [5.4], increasing the coolant temperature actually reduced the probability of preignition. Clearly, in Eq. 5.4, if $n < 1.5$, as intake temperature increases with the intake pressure held constant, the laminar flame thickness will increase at a given crank angle, and the likelihood of flame initiation, and hence preignition, will decrease. Changing the coolant temperature might also affect the likelihood of the oil/fuel droplets coming into the cylinder during the compression stroke and might affect the likelihood of preignition in modern engines.

In [5.22], the consequences of increasing either the intake pressure or the intake temperature have been investigated in greater detail for different fuels. This study highlights the need for better understanding of the pressure and temperature effects on laminar burning velocity (i.e., for establishing n and m values in Eq. 5.4 for different fuels). All else being equal, using cooled EGR will reduce S_{L0} for a given fuel and might be one way to reduce the probability of preignition, as has been demonstrated in [5.31].

5.1.5 Mixture Strength Effects on Preignition

S_{L0} also depends on the mixture strength, with the maximum value occurring at a mixture strength slightly rich of stoichiometric for hydrocarbons and decreasing from this maximum value as the mixture is made leaner or richer [5.26–5.28]. Downs and Theobald [5.29] found that the maximum preignition tendency for different fuels occurred when the mixture strength was around 10% richer than stoichiometric; this is the mixture strength where the laminar burning velocity would be expected to be near maximum for most hydrocarbon fuels. Hamilton et al. [5.30] have studied the effect of mixture strength using a glow plug on the preignition tendency of E85 (85% ethanol and 15% gasoline by volume) in a CFR engine at a fixed condition. With the glow plug power set at 80 watts, they could get preignition occurring in a mixture strength range of 33% rich and 18% lean compared to stoichiometric conditions. Presumably, outside this range, the laminar burning velocity was too low for preignition to occur. Predictably, when the glow plug power was lower, the mixture–strength range in which preignition occurred was narrower [5.30].

5.1.6 Fuel Effects on Preignition

Many previous studies have attempted to identify fuel properties that affect preignition [5.10, 5.14, 5.16, 5.17] without much success. In all these studies a heated wire or a glow plug is used to start the preignition so that the ignition criterion is met. Hence the probability of preignition depends only on the probability of flame initiation. All these studies have found that there is little correlation between the preignition tendency of a fuel and its research or motor octane number (RON or MON).

For liquid fuels, the kinematic viscosity (μ_0/ρ_0) will be similar for different liquid fuels and comparable to that of air because the mixture is predominantly made of air in molar terms [5.22]. Hence, if different fuels are compared for preignition at a comparable pressure and temperature in an engine, as laminar burning velocity increases, preignition should become more likely as the laminar flame thickness decreases (Eq. 5.4). Of course, this would be strictly true if the laminar burning velocity was measured at the same pressure and temperature at which the preignition occurred (i.e., $T_0 = T$ and P0 = P). Failing that, the second and third terms in Eq. 5.4 should be similar for the different fuels being compared. Even though these conditions are not necessarily satisfied, a good correlation between the propensity for preignition of a fuel and laminar burning velocity can still be found from experiments.

Hydrogen, which has a very high laminar burning velocity, is known to be particularly prone to preignition [5.32]. Hydrogen also has a high kinematic viscosity compared to liquid fuels, but the laminar flame thickness will still be much smaller than for other fuels because of its high burning velocity (see Table 1 in [5.22]). PR has been measured for many different single-component hydrocarbon fuels [5.10]. Farrell et al. [5.33] measured the maximum laminar burning velocity, S_{Lmax}, at 3.04 bar (0.304 MPa) (P_0) and 450 K (T_0) for many of these fuels. These data are listed in Table 5.1 (see also [5.22]) with the fuels arranged according to increasing PR; the table also shows RON and MON for these fuels. The PR for ethanol is taken from [5.34] and for the rest of the fuels from [5.10].

PR is plotted against S_{Lmax} from Table 5.1 in Fig. 5.3.

Table 5.1 Preignition Resistance (PR) and Maximum Burning Velocity S_{Lmax} for Different Fuels

Fuel	PR	S_{Lmax}, m/s	RON	MON	Source of RON/ MON
ethanol	-28	0.87	110	91	[5.31]
1-pentene	-21	0.845	90.9	77.1	API
1-hexene	-20	0.835	76.4	74	API
cyclohexane	0	0.78	83	78	API
ethyl benzene	18	0.77	109	97.9	rated
cumene (isopropylbenzene)	19	0.765	113.1	99.3	API
benzene	26	0.84	105	97	API
2-methylbutene	50	0.71	98	82	API
cyclopentane	70	0.782	102.8	85.7	rated
isopentane	75	0.662	93.5	93	API
toluene	93	0.68	117	102	ASTM
p-xylene	95	0.615	113	100.6	rated
iso-octane	100	0.667	100	100	ASTM
o-xylene	120	0.615	105.4	88.8	rated
m-xylene	125	0.56	117	101.3	rated

Notes: PR for ethanol from [5.34], and for the rest of the fuels are from Table 4.9 of [5.10]. S_{Lmax} is from Fig. 15 of Farrell et al. [5.33] and is the maximum laminar burning velocity measured at a pressure (P_0) of 0.304 MPa and a temperature (T_0) of 450 K.

Clearly, as laminar burning velocity increases, PR decreases (i.e. preignition becomes more likely). The correlation is striking, given that S_{Lmax} was measured independently of PR about five decades later, and also given the uncertainties discussed in the previous paragraph.

There is much poorer correlation between PR and RON (Fig. 5.4) or PR and MON (Fig. 5.5).

Ethanol has the highest value of S_{Lmax} at 0.87 m/s, and so we might expect ethanol to be more prone to preignition than the other fuels listed in Table 1. Ethanol was indeed found to be worse than gasoline [5.30] and worse than

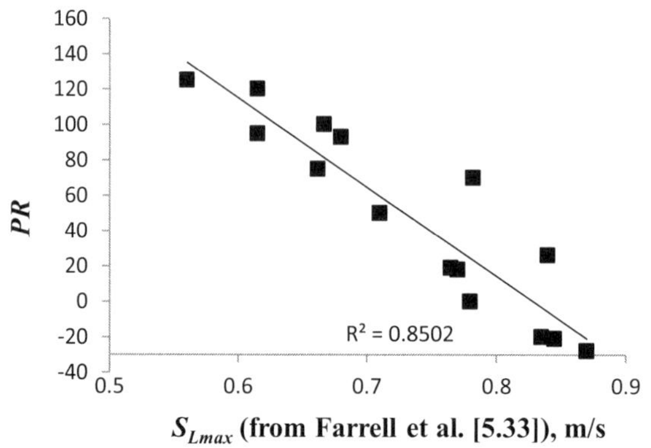

Fig. 5.3. Preignition resistance, PR versus S_{Lmax}. The data used are listed in Table 5.1.

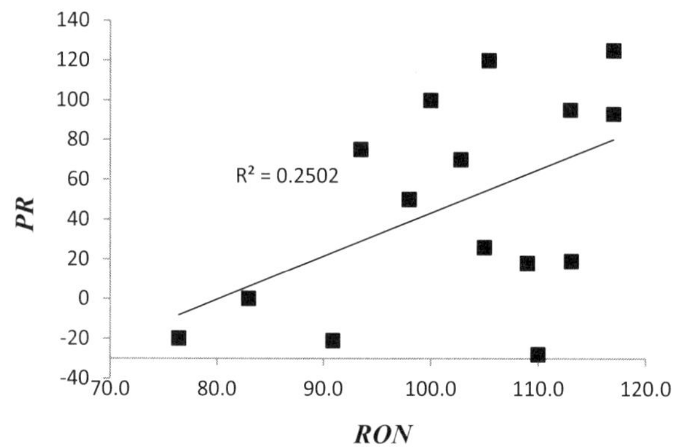

Fig. 5.4. Preignition resistance, PR versus RON.

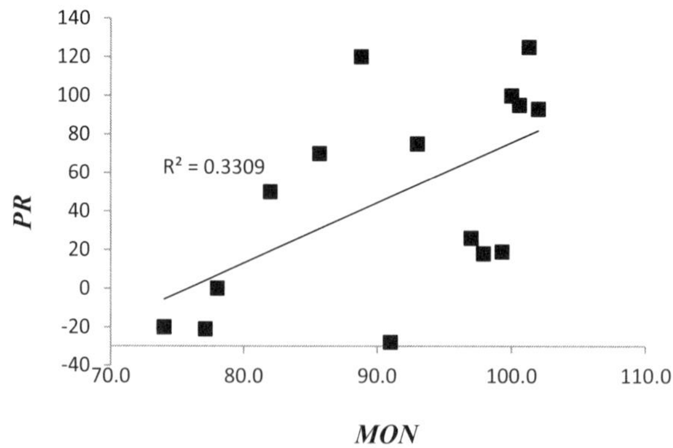

Fig. 5.5. Preignition resistance, PR versus MON.

iso-octane and other gasoline components [5.17] for preignition in port fuel injected engines. However, in a DISI engine, if the initial ignition mechanism is via the burning of oil/fuel droplets, the very high autoignition resistance of ethanol compared to gasoline might make this first step in the preignition process less likely, thereby mitigating the effect of the higher burning velocity of ethanol. Ethanol has a higher heat of vaporization compared to other fuels, and in a DISI engine, it will reduce the mixture temperature more by charge cooling. If the value of n in Eq. 5.4 is higher than 1.5 for ethanol, as the literature suggests [5.27, 5.35], laminar flame thickness will increase with decreasing temperature and make preignition less likely, mitigating the effect of its higher laminar burning velocity.

Thus, preignition is an abnormal combustion phenomenon essentially determined by flame initiation though autoignition, aided by catalytic reactions, that is centered around droplets of lubricant/fuel mixtures might be important in starting the flame in DISI engines. A simple analysis of flame initiation yields Eq. 5.4, which can explain, at least qualitatively, most of the previously unexplained experimental results. All else being equal, a fuel with a higher laminar burning velocity will be more prone to preignition than a fuel of lower laminar burning velocity. However, if preignition is initiated by oil/fuel droplets, mixture (droplet) formation and the ignitability of these droplets are bound to be important in addition to laminar burning velocity.

Preignition will increase the pressure and temperature in the engine well above the values obtained in normal combustion and could lead to very high intensity knock caused by autoignition in the end-gas at high pressures and temperatures.

5.2 Super-Knock

The mixture in the end-gas is never homogeneous even when the fuel and air are fully premixed; there are always gradients of temperature and/or mixture strength. The autoignition that triggers knock is initiated at one or more centers or hot spots. The pressure wave initiated by the autoignition changes as it traverses through inhomogeneity in the hot spot, as has been widely studied [5.36–5.38]. Similar phenomena also occur in homogeneous charge compression ignition (HCCI) engines [5.39, 5.40]. These studies were based on the early work of Zeldovich [5.41] and Oppenheim [5.42]. Hot spots are associated with gradients of reactivity, which lead to gradients of autoignition delay time. When fuel and air are fully premixed, as in an SI or HCCI engine, the reactivity gradient can be assumed to be the temperature gradient. The reaction front will then propagate at a velocity relative to the unburned gas of u_a given by:

$$u_a = (\partial\tau_i / \partial r)^{-1} = (\partial\tau_i / \partial T)^{-1}(\partial T / \partial r)^{-1} \tag{5.5}$$

If it is assumed that can be expressed locally at constant pressure in Arrhenius form by, then:

$$\tau_{\mathrm{i}} = A' \exp\left(\frac{E}{RT}\right) \tag{5.6}$$

where A' is a numerical constant. Then from Eqs. 5.5 and 5.6,

$$u_a = -(RT^2/\tau_i E)(\partial T / \partial r)^{-1} \tag{5.7}$$

The amplitude of the pressure wave generated by autoignition depends on u_a [5.40]. The overpressure above the ambient generated by the propagation of the reaction front through the hot spot can be related, in the absence of shock waves, to a nondimensional resonance parameter ξ defined as:

$$\xi = \frac{a}{u_a} \tag{5.8a}$$

where a is the acoustic speed, $a = \sqrt{\gamma p / \rho}$ [5.43, 5.22]. However, when there are large changes in density at the pressure front, and when chemical energy can be fed into the developing pressure wave at a hot spot, the main energy release occurs during an excitation time τ_e that is orders of magnitude smaller than τ_i. The strength of the pressure pulse is influenced by the rapidity of this energy release [5.44, 5.37]. The development of detonation depends not just on ξ, but also on a dimensionless reactivity parameter ε. If r_0 is the radius of the hot spot, ε is defined as:

$$\varepsilon = \frac{r_0}{a\tau_e} \tag{5.8b}$$

ε is a measure of the hot spot reactivity—it is the number of excitation times fitting into the pressure pulse during its time in the hot spot. The pressure wave generated by the hot spot can couple with the autoignition reaction front when it moves into the unburned mixture at approximately the acoustic speed. The fronts are mutually reinforced to create a damaging pressure spike propagating at high velocity within the hot spot in a developing detonation (DD) [5.41]. This coupling is complex, and a full understanding of a DD requires direct numerical simulation of the developing front. Such simulations were made in [5.40, 5.45] with CO/H_2—air mixtures under different conditions. These fuels were selected because the associated reaction kinetics were relatively well understood. It was found that the results of the computations could be usefully generalized by plotting regime boundary values of ξ and ε as in [5.40] (see Fig. 5.6).

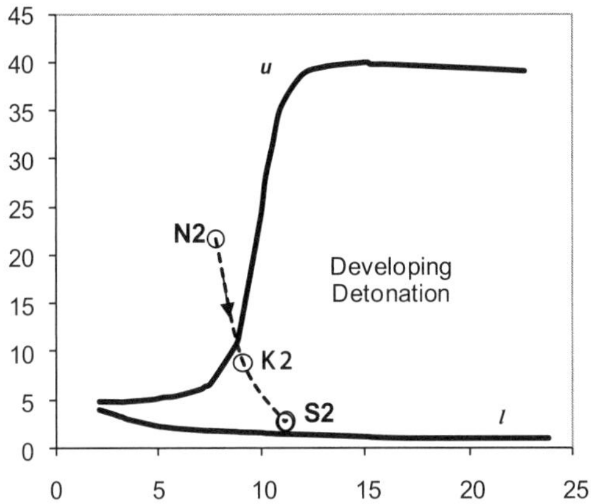

Fig. 5.6. Conditions for the occurrence of DDs in terms of ξ and ε.

Chemical resonance corresponds to $\xi = 1$. There is a peninsula in Fig. 5.6 with an upper limit ξ_u and a lower limit ξ_l within which detonations can begin to develop within the hot spot. The analysis in [5.40, 5.45] identified the modes of propagation within the hot spot as follows:

1. $\xi = 0$, thermal explosion.
2. $0 < \xi < \xi_l$, supersonic autoignitive deflagration. The autoignitive wave is ahead of acoustic wave.
3. $\xi_l \leq \xi < \xi_u$, developing and developed detonation.
4. $\xi_l \leq \xi < aS_l^{-1}$, subsonic autoignitive deflagration where S_l is the laminar burning velocity.
5. $\xi \geq aS_l^{-1}$, laminar burning deflagration at the laminar burning velocity.

We can assess the mode of propagation of the pressure wave for a cycle that shows autoignition, whether in a knocking SI engine or an HCCI engine, using Fig. 5.6 if we can estimate ξ and ε. Such an assessment has been made for the cycles shown in Fig. 5.1 in [5.7]. The pressure and temperature at the start of autoignition, P_{ai} and T_{ai}, estimated from the pressure curve, are established. The acoustic velocity a can be calculated from the temperature at the start of autoignition. The temperature gradient in Eq. 5.5 has to be assumed.

Experimental techniques currently do not have the spatial resolution or the accuracy to measure this accurately; a value of –2 K/mm was assumed in [5.7].

To calculate the first term on the right hand side of Eq. 5.5, a chemical kinetic model was employed to estimate the ignition delay parameters. A surrogate fuel, made up of toluene (9% vol), iso-octane (62% vol), and n-heptane (29% vol), that behaves very similarly in terms of knock to the actual fuel was used. Ignition delays were calculated using the chemical kinetic model described in [5.46] at different pressures and temperatures—the results of these calculations are shown in Fig. 5.7a. By multiple regression, with T and P as independent variables, and $\ln(\tau_i)$ as the dependent variable, a single equation was fitted to these data:

$$\tau_i = 83.1\exp\left(\frac{6626}{T}\right)P^{-1.21} \tag{5.9}$$

with P in bar and T in K. Figure 5.7b shows that Eq. 5.9 is a very good approximation for τ_i over the range of P and T considered in Fig. 5.7a.

Using Eq. 5.9, τ_i can be calculated as a function of crank angle for the cycles shown in Fig. 5.1 since the pressure is known and the temperature of the unburned gas ahead of the flame front can be estimated [5.7] at each crank angle. This also allows the Livengood-Wu integral I to be assessed for each cycle (see Chapter 4). Autoignition, and hence knock, is deemed to occur when I reaches unity. The crank angle position for the onset of knock estimated using this method was in excellent agreement with the observed value for Cycles S2 and K2 in Fig. 5.1 as well as for another engine cycle from another engine using a comparable fuel [5.7]. Hence, Eq. 5.9 is a very good predictor of τ_i for a European premium gasoline.

Then u_a could be estimated, for a fixed pressure, using Eq. 5.7 so that ξ could be calculated. The excitation time τ_e defined as the time from 5% of maximum to the point of maximum heat release rate, was also calculated using the chemical kinetic mechanism [5.46], but using the smallest time increment available in CHEMKIN [5.47], the chemistry simulation software that was used, for different pressures and temperatures [5.7]. Again, a single equation, Eq. 5.10, could be fitted to these data.

$$\tau_e = 3.07\exp\left(\frac{799}{T}\right)P^{-0.456} \tag{5.10}$$

The radius of the hot spot r_0 has not been experimentally determined, but it was assumed to be 5 mm in [5.7] so that the reactivity parameter ε (Eq. 5.8b)

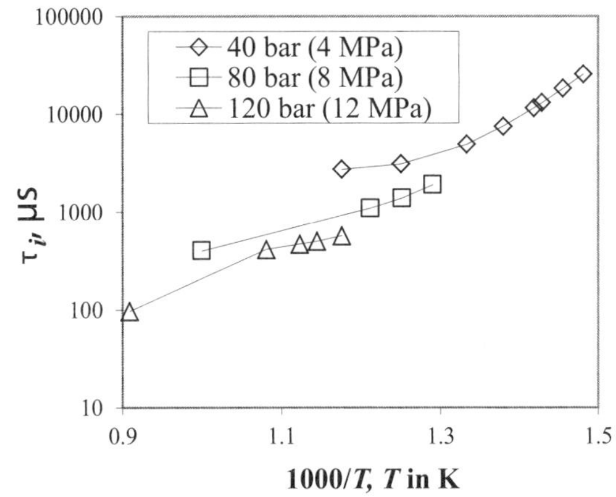

Fig. 5.7a. Ignition delay τ_i for a surrogate gasoline of RON 98 and MON 91 (62% vol iso-octane, 29% vol toluene, and 9% vol n-heptane) using the kinetic model from [5.46].

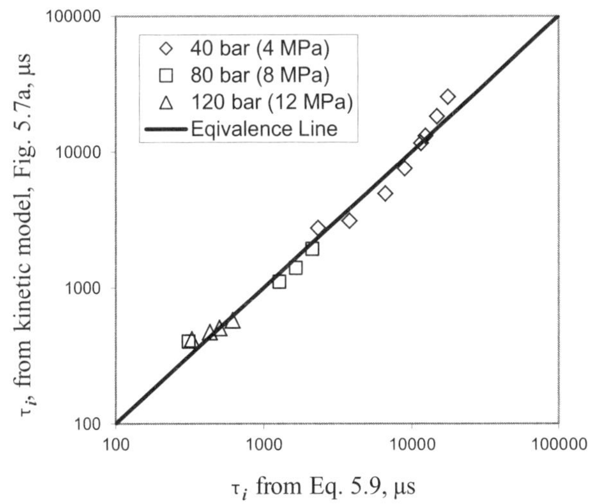

Fig. 5.7b. Comparison of τ_i from Fig. 5.7a with τ_i calculated from Eq. 5.9.

could be estimated. The results are shown in Table 5.2 and plotted in Fig. 5.6: P_{ai} and T_{ai} are the pressure and temperature at the start of autoignition estimated from the pressure traces.

Also shown in Table 5.2 is the estimation of these parameters at –19 CAD BTDC, the point where preignition starts in Cycle S2 in Fig. 5.1 using the mean temperature of 620 K at this point in the cycle. It can be seen that u_a is only 0.39 m/s at this point and thus almost certainly lower than the laminar burning

Table 5.2 Estimated Parameters for the Three Cycles Shown in Fig. 5.1 Assuming ($\partial T/\partial r$) = 2 K/mm and r_0 = 5 mm

Cycle	P_{ai}, bar (MPa)	T_{ai}, K	a, m/s	τ_e, μs	$a\tau_e$ mm	τ_i, μs	$\partial\tau_i/\partial T$ μs/K	u_a, m/s	ε	ξ
S2	128 (12.8)	926	577	0.78	0.45	300	2.3	216	11.1	2.7
K2	90 (9.0)	850	555	0.99	0.55	870	8.0	62.7	9.1	8.8
N2	70 (7.0)	800	540	1.18	0.64	1924	19.9	25	7.8	21.6
S2 @ −19 CAD	25 (2.5)	620	479	2.54	1.22	74036	1276	0.39	4.1	1222

velocity of the bulk fuel/air mixture. Hence, any hot spot at this pressure and temperature would give rise to a laminar flame (i.e., the condition for mode 5 in Section 5.2 would be satisfied), though the hot spot itself cannot arise from the autoignition of the fuel/air mixture in preignition events. Once the laminar flame is established, it would be affected by turbulence surrounding it and develop into a turbulent flame.

Cycle N2 from Fig. 5.1 is well outside the DD peninsula in Fig. 5.6, while Cycle S2 falls well within this peninsula. Another super-knock cycle from a different engine (not considered here) also falls well within the DD peninsula [5.7]. A similar analysis in yet another engine [5.8] also shows that very high KI cycles following a preignition could be the result of a DD. As the pressure at the start of knock increases, ε increases and ξ decreases [5.7, 5.22], and the severity of knock increases [5.22]. The much higher KI (super-knock) in Cycle S2 compared to Cycle N2 with a comparable fraction of unburned mixture at the start of knock is most likely the result of a DD in Cycle S2. There could, of course, be multiple hot spots. Under such circumstances, the coupling of reaction fronts and shock waves could create further DDs [5.48] and cause super-knock. Further discussion about the modes of autoignition in this context can be found in [5.22]. It should also be remembered that a preignition does not necessarily lead to a DD and super-knock [5.8]. Thus the probability of super-knock will be higher when the pressure is higher and temperature at which autoignition occurs in the end gas is higher; this could be brought about by preignition earlier in the cycle.

The analysis considered above, of the modes of propagation of the reaction front originating from a hot spot, does not include turbulence. Moreover, the temperature gradient in Eq. 5.5 and the radius of the hot spot (r_0) in Eq. 5.8b have to be assumed. Peters et al. [5.49] pointed out that under turbulent conditions these quantities can only be described in statistical terms. They have used a refinement of turbulence theory to estimate the joint probability of the temperature gradient and the length scale l over which the temperature gradient exists; l is taken as a measure of r_0. They can then estimate the probability of DD and thus super-knock occurring in an engine running at a given pressure and temperature regime on a fuel whose autoignition behavior can be modeled [5.48].

Both ξ and ε and, consequently, the probability of DD depend on parameters defining the autoignition resistance and, hence, on the octane numbers of the fuel. Bradley (Fig. 3 in [5.37]) showed that considering primary reference fuels (PRFs), ξ is lower for, say, PRF 90 compared to iso-octane, and therefore, with all else the same, we should expect DD to be more likely for PRF 90. Peters et al. [5.48] found that the probability of DD is higher for PRF 85 compared to iso-octane. One would expect fuels like ethanol and hydrogen, which are more resistant to autoignition compared to iso-octane, to be less likely to give rise to DD and super-knock compared to iso-octane, all else being equal.

Preignition enables much higher pressures to be reached in the cylinder (compared to normal combustion initiated by the spark) so that knock becomes more likely. If preignition is avoided, any kind of knock, including DD, will be avoided. So the probability of super-knock in an engine depends on two probabilities: the probability P_{PI} of a preignition and the probability P_{DD} of a DD should autoignition occur.

Peters et al. developed a framework for estimating P_{DD} [5.48]; Kalghatgi and Bradley [5.22] argued that preignition depends on establishing an initial laminar flame, and all else being equal, P_{PI} increases as laminar burning velocity increases. Preignition can be detected by the increase in cylinder pressure above the compression pressure before the spark plug fires. Experimentally, it has been established that fuels with high laminar burning velocities like ethanol and hydrogen have high P_{PI}. Fuels could be ranked for P_{DD} by running the engine at a more advanced spark timing than normal to induce knock and then comparing the KI of cycles when autoignition is initiated at the same pressure and temperature for different fuels. Such an experiment is likely to show that ethanol (or hydrogen) will have lower P_{DD} compared to, say, iso-octane. On the other hand, the probability of super-knock could be estimated by running the engine at a fixed condition (e.g., with the normal spark ignition timing which avoids knock in the absence of preignition) and then establishing the frequency of occurrence of knock cycles of very high intensity (say, above 80 bar [8 MPa]). In such experiments, it is possible that, compared to iso-octane, ethanol is as or less likely to suffer from super-knock because the higher P_{PI} of ethanol is cancelled out by its lower P_{DD}.

5.3 Summary and Concluding Remarks

Preignition, where a flame front is initiated before the spark fires, is a potential problem in turbocharged engines. It causes the pressure and temperature ahead of the advancing flame front to rise more rapidly when compared to normal spark timing. In many cases, extremely intense knock, loosely referred to as super-knock, can occur following preignition. These abnormal combustion phenomena could seriously limit the development of efficient SI engines.

For preignition to occur, an ignition criterion has to be satisfied whereby the local temperature is increased till a thermal runaway is initiated by some external heat source before the spark fires. This initial thermal runaway, the source of preignition, cannot be because of the autoignition of the bulk fuel/air mixture since, the ignition delays are too low when preignition is normally observed to occur during the compression stroke before the spark plug fires. This source can be hot spots on internal surfaces of the engine, or in modern engines, most probably combustion of oil/fuel droplets brought about by autoignition and catalytic reactions. However, following ignition, for a stable flame front to be set up, an initiation criterion has to be met. The probability of the ignition criterion being met depends on many stochastic processes in the engine; the probability of the initiation criterion being met increases with decreasing laminar flame thickness, which, in turn, is inversely proportional to the laminar burning velocity (Eq. 5.4). As the pressure increases, laminar flame thickness decreases, and hence the probability of preignition occurring increases, all else being equal.

In many experiments on preignition, the ignition criterion is met by using an electrically heated wire or plug, so that the probability of preignition depends only on the initiation criterion being met. For a given engine-operating condition, a fuel with a larger laminar burning velocity is more likely to cause preignition. This conclusion is borne out by experimental results in the literature. The likelihood of preignition can be reduced by reducing the probabilities of either the ignition or the flame initiation criteria being met. For instance, if preignition is initiated by the combustion of oil/fuel droplets, preignition can be reduced by reducing fuel impingement on the walls, by increasing the crevice volume in the top land of the piston, or by removing metallic additives from the oil that appear to promote catalytic reactions and hence combustion of the oil/fuel droplets. Perhaps making the lubricant more resistant to autoignition by adding antiknock additives might also reduce the probability of the initial ignition event leading to preignition. Using cooled EGR would reduce preignition by reducing the laminar burning velocity and making flame initiation less likely.

Preignition increases the pressure and temperature in the end-gas, ahead of the advancing flame front very significantly compared to a normal cycle without preignition and autoignition leading to knock could occur. As the pressure and temperature at the onset of knock increase, KI will increase and could lead to DD, which is capable of producing the extremely high KI informally described as super-knock. The severity of such autoignition events could also be increased by the occurrence of multiple hot spots. However, super-knock can be avoided in practical engines if preignition can be avoided.

5.4 References

5.1 Dahnz, C., Han, K-H, Spicher, U., Magar, M., Schiessel, R., and Maas, U. 2010. "Investigation on Pre-ignition in Highly Supercharged Engines." SAE Paper No. 2010-01-0355. SAE International, Warrendale PA.

5.2 Dahnz, C., and Spicher, U. 2010. "Irregular Combustion in Supercharged Spark Ignition Engines—Pre-Ignition and Other Phenomena." *International Journal of Engine Research* 11(6):485–498.

5.3 Manz, P-W, Daniel, M., Jippa, K-N, and Willand, J. 2008. "Pre-ignition in Highly Turbo-charged Engines. Analysis Procedure and Results." Paper presented at 8th International Symposium on Internal Combustion Diagnostics, Baden-Baden.

5.4 Zahdeh, A., Rothenberger, P., Nguyen, W., Anbarasu, M., Schmuck-Soldan, S., Schaefer, J., and Goebel, T. 2011. "Fundamental Approach to Investigate Pre-ignition in Boosted SI Engines." SAE Paper No. 2011-01-0340. SAE International, Warrendale PA.

5.5 Han, K-M, Sauter, W., and Spicher, U. 2008. "3-D Visualization of Spark-Ignition Combustion: Practical Examples of Flame Propagation, Abnormal Combustion and Controlled Compression Ignition." Paper presented at the 8th International Symposium on Internal Combustion Diagnostics, Baden-Baden.

5.6 Amann, M., Mehta, D., and Alger, T. 2011. "Engine Operating Condition and Gasoline Composition Effects on Low-Speed Pre-ignition in High Performance Spark Ignited Engines." SAE Paper No. 2011-01-0342. SAE International, Warrendale PA.

5.7 Kalghatgi, G.T., Bradley, D., Andrae, J., and Harrison, A.J. 2009. "The Nature of 'Super-Knock' and Its Origins in SI Engines." In *Conference on Internal Combustion Engines: Performance, Fuel Economy and Emissions*, edited by Institution of Mechanical Engineers, Chandos, London.

5.8 Rudloff, J., Zaccardi, J-M, Richard, S., and Anderlohr, J.M. 2013. "Analysis of Pre-ignition in Highly Charged SI Engines: Emphasis on the Auto-Ignition Mode." *Proceedings of the Combustion Institute* 32(2):2959–2967.

5.9 Hüsler, T., Grünfeld, G., Brands, T., Günther, M., and Pischinger, S. 2013. "Optical Investigation of Pre-ignition in a Highly Boosted SI Engine Using Bio-Fuels." SAE Paper No. 2013-01-1636. SAE International, Warrendale PA.

5.10 Ricardo, H.R., and Hempson, J.G.G. 1972. *The High-Speed Internal Combustion Engine*. Blackie and Son, London.

5.11 Winch, R.F., and Mayes, F.M. 1953. "A Method of Identifying Pre-Ignition." *SAE Transactions* 61:453–468. SAE International, Warrendale PA.

5.12 Hirschler, D.A., McCullough, J.D., and Hall, C.A. 1954. "Deposit-Induced Ignition—Evaluation in a Laboratory Engine." *SAE Transactions* 62:41–49. SAE International, Warrendale PA.

5.13 Melby, A.O., Diggs, D.R., and Sturgis, B.M. 1954. "An Investigation of Preignition in Engines." *SAE Transactions* 62:32–38. SAE International, Warrendale PA.

5.14 Sturgis, B.M., Cantwell, E.N., Morris, W., and Schultz, D.L. 1954. "The Preignition Resistance Of Fuels." *Transactions of the API, Division of Refining* 34:256–269.

5.15 Pahnke, A.J. 1963. "Surface Ignition; Factors Affecting Its Occurrence in Engines." SAE Paper No. 630489. SAE International, Warrendale PA.

5.16 Guibet, J.C., and Duval, A. 1972. "New Aspects of Preignition in European Automotive Engines." SAE Paper No. 720114. SAE International, Warrendale PA.

5.17 Menrad, H., Haselhorst, M., and Erwig, W. 1982. "Pre-ignition and Knock Behaviour of Alcohol Fuels." SAE Paper No. 821210. SAE International, Warrendale PA.

5.18 Suga, T., Kitajima, S., and Fujii, I. 1989. "Pre-ignition Phenomena of Methanol (M85) Fuel by the Post-Ignition Technique." SAE Paper No. 892061. SAE International, Warrendale PA.

5.19 Swain, M.R., Blanco, J.A., and Swain, M.N. 1989. "Abnormal Combustion in a Methanol Fuelled Engine." SAE Paper No. 892162. SAE International, Warrendale PA.

5.20 Lewis, B., and von Elbe, G. 1961. *Combustion, Flames and Explosions of Gases*. Academic Press, New York.

5.21 Joulin, G. 1985. "Point Source Initiation of Lean Spherical Flames of Light Reactants: An Asymptotic Theory." *Combustion Science and Technology* 43:99–113.

5.22 Kalghatgi, G.T., and Bradley, D. 2012. "Pre-ignition and 'Super-Knock' in Turbocharged Spark Ignition (SI) Engines." *International Journal of Engine Research* 13(4):399–414.

5.23 Kalghatgi, G.T. 1987. "Spark Ignition, Early Flame Development and Cyclic Variation in I.C. Engines." SAE Paper No. 870163. SAE International, Warrendale PA.

5.24 Zeldovich, Y. B., Barenblatt, G.I., Librovich, V.B., and Makhviladze, G.M. 1985. *Combustions and Explosions*. Consultants Bureau.

5.25 He, L., and Clavin, P. 1993." Premixed Hydrogen-Oxygen Flames. Part II: Quasi-Isobaric Ignition Near the Flammability Limits." *Combustion and Flame* 93:408–420.

5.26 Metghalchi, H., and Keck, J.C. 1982. "Burning Velocities of Mixtures of Air with Methanol, Isooctane and Indolene at High Pressure and Temperature." *Combustion and Flame* 48:191–210.

5.27 Gulder, O.L. 1982. "Laminar Burning Velocities of Methanol, Ethanol and Isooctane-Air Mixtures." *Nineteenth Symposium (International) on Combustion*. 275–281. The Combustion Institute.

5.28 Bradley, D., Hicks, R.A., Lawes, M., Sheppard, C.G.W., and Woolley, R. 1998. "the Measurement of Laminar Burning Velocities and Markstein Numbers for Iso-octane—Air and Iso-octane-N-heptane-Air Mixtures at Elevated Temperatures and Pressures in an Explosion Bomb." *Combustion and Flame* 115:126–144.

5.29 Downs, D., and Theobald, F.B. 1963. "the Effect of Fuel Characteristics and Engine Operating Conditions on Pre-ignition." *Proceedings of the Institution of Mechanical Engineers (Auto Division)* 178(2A):89–102.

5.30 Hamilton, L., Caton, P., and Cowart, J. 2008. "Pre-ignition Characteristics of Ethanol and E85 in a Spark Ignition Engine." SAE Paper No. 2008-01-0231. SAE International, Warrendale PA.

5.31 Amann, M., Alger, T., and Mehta, D., 2011. "The Effect of EGR on Low-Speed Pre-ignition in Boosted SI Engines." SAE Paper No. 2011-01-0039. SAE International, Warrendale PA.

5.32 Verhelst, S., Verstraeten, S., and Sierens, R. 2007. "A Comprehensive Overview of Hydrogen Engine Design Features." *Proceedings of the Institution of Mechanical Engineers, Part D: Journal of Automobile Engineering* 221(8):911–920.

5.33 Farrell, J.T., Johnson, R.J., and Androulakis, I.P. 2004. "Molecular Structure Effects on Laminar Burning Velocities At Elevated Temperature and Pressure." SAE Paper No. 2004-01-2936. SAE International, Warrendale PA.

5.34 Thring, R.H. 1983. "Alternative Fuels for Spark Ignition Engines." SAE Paper No. 831685. SAE International, Warrendale PA.

5.35 Bradley, D., Lawes, M., and Mansour, M. S. 2009. "Explosion Bomb Measurements of Ethanol-Air Laminar Flame Characteristics at Pressures Up to 1.4 MPa." *Combustion and Flame* 156:1462–1470.

5.36 König, G., Maly, R. R., Bradley, D., Lau A. K. C., and Sheppard C. G. W. 1990. "Role of Exothermic Centres on Knock Initiation and Knock Damage." SAE Paper No. 902136. Also SAE Transactions. SAE International, Warrendale PA.

5.37 Bradley, D. 2012. "Autoignitions and Detonations in Engines and Ducts." *Philosophical Transactions R of the Royal Society A* 370:689–714.

5.38 Pan, J., Sheppard, C.G.W., Tindall, A., Berzins, M., Pennington, S.V., and Ware, G.M. 1998. "End Gas Inhomogeneity, Auto-Ignition and Knock." SAE Paper No. 982616, SAE International, Warrendale PA.

5.39 Sheppard, C.G.W, Tolegano, S., and Wooley, R. 2002. "On the Nature of Auto-ignition Leading to Knock in HCCI Engines." SAE Paper No. 2002-01-2831. SAE International, Warrendale PA.

5.40 Bradley, D., Morley, C., Gu, X.J., and Emerson, D.R. 2002. "Amplified Pressure Waves during Auto-ignition: Relevance to CAI Engines." SAE Paper No. 2002-01-2868. SAE International, Warrendale PA.

5.41 Zeldovich, Y. B. 1980. "Regime Classification of an Exothermic Reaction with Non-uniform Initial Conditions." *Combustion and Flame* 39:211.

5.42 Oppenheim, A.K. 1985. "Dynamic Features of Combustion." *Philosophical Transactions R of the Royal Society A* 315:471–508.

5.43 Bradley, D., and Kalghatgi, G.T. 2009. "Influence of Autoignition Delay Time Characteristics on Pressure Fluctuations and Knock in Reciprocating Engines." *Combustion and Flame* 156:2307–2318.

5.44 Lutz, A. E., Kee, R. J., Miller, J. A., Dwyer, H. A., and Oppenheim, A. K. 1988. "Dynamic Effects of Autoignition Centers for Hydrogen and $C_{1,2}$-Hydrocarbon Fuels." *Proceedings of the Combustion Institute* 22:1683–1693. The Combustion Institute.

5.45 Gu, X. J., Emerson, D. R., and Bradley, D. 2003. "Modes of Reaction Front Propagation from Hot Spots." *Combustion and Flame* 133:63–74.

5.46 Andrae, J., P. Björnbom, R.F. Cracknell, G.T. Kalghatgi. 2007. "Auto-ignition of Toluene Reference Fuels at High Pressures Modelled with Detailed Chemical Kinetics." *Combustion and Flame* 149:2–24.

5.47 "CHEMKIN." Reaction Design. http://www.reactiondesign.com/products/open/chemkin.html. Accessed February 6, 2013.

5.48 Pan, J., and Sheppard, C.G.W. 1994. "A Theoretical and Experimental Study of the Modes of End Gas Autoignition Leading to Knock in SI Engines." SAE Paper No. 942060. SAE International, Warrendale PA.

5.49 Peters, N., Kerschgens, B., and Paczko, G. 2013. "Super-Knock Using a Refined Theory of Turbulence." SAE Paper No. 2013-01-1109. SAE International, Warrendale PA.

Chapter 6
Fuel Effects on Compression Ignition Combustion—Is Low-Octane Gasoline the Best Fuel for Advanced Diesel Engines?

Practical compression ignition (CI) engines are diesel engines running on conventional diesel fuels. In diesel engines, diesel fuel is injected into hot and high-pressure gas near TDC (top dead center), near the end of the compression stroke. Fuel and air are not fully premixed, as in spark ignition (SI) or HCCI engines, and combustion is initiated by autoignition as the fuel vaporizes, mixes, and reacts with oxygen in the engine cylinder. In contrast to HCCI engines, combustion phasing is controlled by the timing of the final fuel injection before combustion. The amount of the fuel injected depends on the injection pressure, the area of the injector hole, the number of injector holes, and the duration of the injection. Diesel engines are much more efficient than SI engines for three reasons [6.1]:

1. In contrast to SI engines, diesel engines can run at part load by reducing the amount of fuel injected rather than by controlling the mass of air trapped in the cylinder by the use of a throttle. Throttling losses are the main source of inefficiency of SI engines at part load, and most engines run predominantly at part load in practice.
2. Diesel engines do not suffer from knock at high loads and hence can have higher compression ratios compared to SI engines.

3. During the compression stroke, diesel engines compress mostly air rather than a mixture of fuel and air, which brings them nearer to the ideal cycle efficiency.

However, diesel engines running on diesel fuels suffer from high engine-out nitrogen oxides (NOx) and soot/smoke/particulate emissions that are difficult to control through aftertreatment. Conventional diesel fuels are very prone to autoignition—their cetane numbers (CNs) are greater than ~40—and combustion starts very soon after fuel is first injected into the cylinder. It is often not possible to complete the final injection of fuel, particularly at higher loads, before combustion starts with conventional diesel fuels, and the diesel fuel will burn in a quasi-steady jet diffusion flame, as described by Dec [6.2]. The fuel and air first react in a fuel-rich mixture, with the equivalence ratio φ around 4, leading to soot formation. Then this rich mixture burns out in a high-temperature diffusion flame at the jet periphery, leading to the formation of NOx [6.2, 6.3]. Soot formation can be minimized by ensuring that the equivalence ratio φ of the mixture packets where combustion occurs is not greater than $\varphi > \sim 2$ or $\lambda < \sim 0.5$, where $\lambda = (1/\varphi)$, is the normalized air/fuel ratio (see Fig. 6.1 and [6.4, 6.5]).

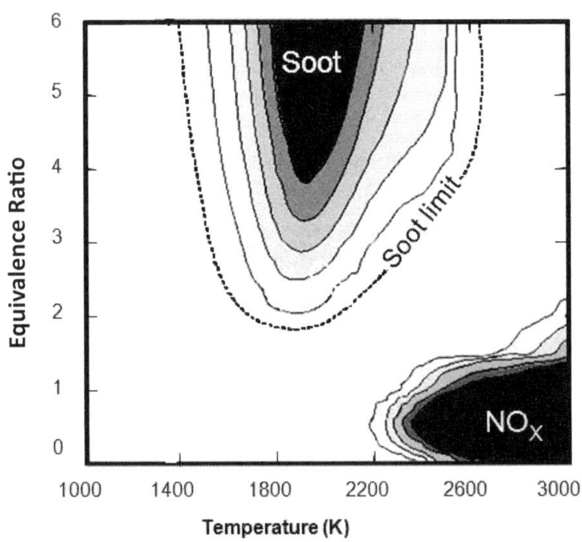

Fig. 6.1. Equivalence ratio and temperature effects on soot and NOx formation, after [6.4, 6.5]. Soot formation can be avoided if $\varphi < \sim 2$, and NOx formation can be minimized by reducing the combustion temperature.

NOx formation can be minimized if the combustion temperatures are kept below about 2200 K [6.4], which can be achieved by using high levels of exhaust gas recirculation (EGR). NOx formation can also be avoided by getting the combustion to occur when the mixture is very lean, with λ much greater than 1, but this is not possible in conventional diesel engine combustion where the mixture starts rich and combustion occurs before the fuel is sufficiently mixed with oxygen. Diesel engines running on diesel fuel cannot avoid soot and NOx formation regions in Fig. 6.1 at most operating conditions. Engine-out soot reflects the balance between the formation and the in-cylinder oxidation of the soot that is formed. Both the temperature and oxygen content in the cylinder decrease if the EGR level is increased to control NOx; however, oxidation of soot decreases and engine-out soot or particulate matter (PM) increases. Therefore, it is very difficult to control both soot and NOx simultaneously in diesel engines using conventional diesel fuels.

EGR could be used to reduce combustion temperatures and thus engine-out NOx if soot formation is avoided in the first place. However, this requires that the equivalence ratio is below about 2 before combustion starts everywhere in the cylinder, and it is not possible if the fuel is still being injected after combustion has started. Injection pressures in diesel engines have increased significantly over the years to avoid soot formation by accelerating the mixing of the fuel with air before combustion starts. However, such high-pressure injection systems are expensive. Regulations to control NOx and PM have become progressively stringent across the world [6.6, 6.7] and will require the use of NOx and PM aftertreatment systems in spite of the high injection pressures if conventional diesel fuel is to be used in diesel engines. This will further increase the cost of the engine.

Conventional diesel engines have become expensive and complicated primarily because they try to control NOx and soot while simultaneously using conventional diesel fuel that ignites early in the mixing process. Low-NOx and low-soot operation of CI engines becomes much easier if the fuel used is more resistant to autoignition than conventional diesel fuel. Such combustion usually increases CO and HC emissions because of the lower temperatures, shifting the focus of aftertreatment to CO and HC control if the engine-out NOx and soot were to be reduced to sufficiently low levels. Reduction of CO and HC through an oxidation catalyst would be easier than NOx aftertreatment when there is excess oxygen in the exhaust and if the exhaust temperature is high enough. There is great potential for developing CI engines that are as efficient and clean but less complicated and less expensive compared to current diesel engines using gasoline-like fuels with increased autoignition delay times. There are differences in emissions regulations and other requirements between heavy-duty (HD) engines used in, say, trucks and light-duty

(LD) engines used in passenger cars. For instance, as compared to LD engines, HD engines are larger and their maximum speeds are lower. NOx and soot need to be controlled over an emissions cycle in LD engines so that at a given engine speed they do not need to control NOx and soot at all loads. In contrast, HD engines, particularly in the United States, need to meet stringent NOx and soot requirements even at full load at any given speed. More expensive technology such as high-pressure injection could perhaps be more easily introduced into HD engines, given that they are in vehicles that are more expensive. This chapter will discuss these issues.

Illustrative experimental results in the following discussion are taken from [6.8–6.16], which describe experiments in two different single-cylinder engines and a European 2.7 L V-6 engine using a variety of different fuels. The emissions shown are averages sampled over a time interval, usually 60 seconds, and the pressure-based parameters are averages from 100 or 250 cycles collected once the engine operation was stabilized; further experimental details can be found in the relevant references. Some of the properties (Chapter 2) of some these fuels are listed in Table 6.1, and the volatility characteristics are shown in Fig. 6.2.

These fuels span the gasoline and diesel boiling range, as well as the autoignition quality range. n-heptane has a CN of 54—similar to a European diesel—but is very much more volatile. Fuel D4 is in the diesel boiling range, but has a measured derived CN (DCN; Chap. 2) of about 22 and behaves, in autoignition terms, like a gasoline (as will be discussed in Section 6.3). Fuel D4 was made by adding high levels of an aromatic solvent to a European diesel fuel. The terms *smoke*, *soot*, and *particulate matter* (PM) are used interchangeably in this chapter. The smoke measurements discussed were all made with an AVL smoke meter, in which a controlled volume of exhaust is drawn through a filter paper, and the smoke level is determined by the change in the reflectance of the paper.

Efficiency, emissions, and maximum heat release rates are affected by combustion phasing, which depends on engine-operating parameters such as injection timing, pressure and strategy, engine speed, and the ignition delay (ID). Combustion phasing is commonly specified by CAx, where CAx is the crank angle position averaged over a number of engine cycles at which x% of the total heat release occurs. Another measure that has been used to characterize the autoignition behavior of a fuel in diesel engines is the combustion delay, CD = CA50 – SOI [6.8–6.14] where SOI is the start of injection.

Table 6.1 Properties of Some of the Fuels Considered in Chapter 6

	D1 Europe Diesel	Swedish MK1	Diesel 39 CN	D2 39.6 CN	n-Hep	Kero F1	ULG 95	D4	84 RON G4	91 RON G5
Density, g/cc	0.833	0.810	0.810	0.852	0.632	0.738	0.726	0.87	0.736	0.738
DCN	56.2	54	39	39.6	~ 55	44.5	~15*	22.5	~19*	~16*
Aromatics v%	25.2	3	34	42.5	0	0	28.5	75.3	26.5	29.8
LHV*** MJ/kg	42.9	43.8	43.5	42.6	44.6	44.3	43.2	42.1	43.2	43.1
Sulfur ppm	7.	10	12	10	0	< 10	< 10	< 10	< 10	<10
Lubricity, microns**	285	280	290	332	270	225	~260	~250	241	254
RON					0		94.7		84	90.7
MON					0		85.9		78	81.8
C					7	7.06	6.7		6.68	6.63
H					16	15.3	12.8		12.5	12.2
H/C	1.97				2.29	2.17	1.91	1.59	1.87	1.840

Notes: The fuels span both the gasoline and diesel boiling range and autoignition quality range.
*Derived cetane number (DCN) is estimated from DCN = 54.6 − 0.42*RON [6.17]
**All fuels had a lubricity additive—see Chapter 2
*** LHV = lower heating value

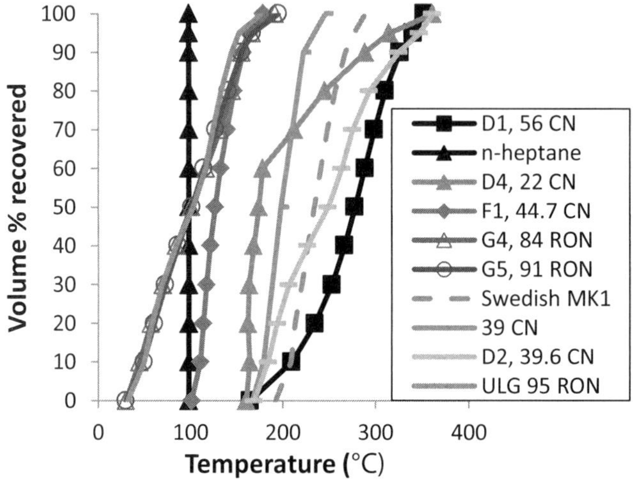

Fig. 6.2. Volume percentage recovered versus temperature in the ASTM volatility test for the fuels in Table 6.1.

6.1 Premixed Compression Ignition

If the final injection event is completed before combustion begins, there will be a distribution of mixture strengths just prior to combustion. All the combustion occurs in a premixed mode, that is, in mixture packets that have been mixed with oxygen to different degrees. The fuel injected first will have had longer time to mix and thus will have a lower φ than the fuel that is injected at the end of the injection pulse. Premixed compression ignition (PCI) combustion was defined to occur in [6.15] when combustion starts after the final fuel injection is over and the engine-out smoke is very low, below 0.05 FSN (filter smoke number). We will use this definition.

A relevant time constant is the ignition dwell, IDW = SOC − EOI where SOC and EOI are the crank angle at the start of combustion and at the end of injection [e.g., 6.14]. IDW is sometimes described as the mixing period. PCI combustion is characterized by positive IDW. If IDW is negative, a quasi-steady lifted jet diffusion flame will result, because combustion will start before the injection event is over. It is impossible to avoid rich equivalence ratios where soot is formed unless IDW is positive. However, IDW needs to be sufficiently positive to allow enough time for the last portion of the injected fuel to mix well enough with oxygen to avoid soot-forming equivalence ratios.

Unlike in HCCI combustion, in PCI combustion, fuel and air are not *fully* mixed, and combustion phasing can be controlled by the timing of the final fuel injection before combustion. Whether PCI combustion is possible depends on both the duration of the final injection event and the autoignition delay time of the mixture. The duration of the injection event with a given injector depends on the amount of fuel that needs to be injected (i.e., on the engine load and on the injection pressure). The resistance to autoignition of the fuel is reflected in the ignition delay (ID), which is normally defined in diesel engines as the difference between the SOI and SOC [6.18]. The ID also depends on engine-operating parameters such as injection timing, quantity, drop size, injection rate, charge temperature and pressure, engine speed, and EGR levels [6.18]. The maximum load at which PCI is possible at any given speed with a given injector and fuel is determined by the injection pressure: The larger the injection pressure, the larger the amount of fuel that could be injected before combustion starts. Fig. 6.3 is a rough, qualitative sketch of the variation of NOx, HC, and CO with mixture strength at a given temperature.

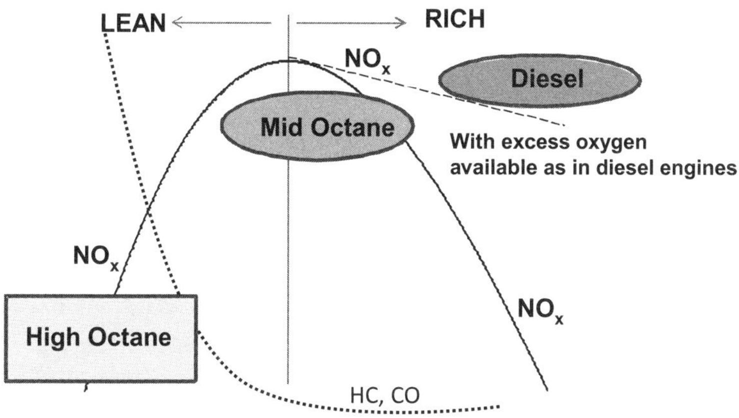

Fig. 6.3. Qualitative sketch of variation of nitrogen oxides (NOx), carbon monoxide (CO), and unburned hydrocarbons (HC) with mixture strength at a given temperature.

We can assume that combustion will take place in mixture packets distributed inside the cylinder. The mixture packets will cover different ranges of mixture strengths before combustion in PCI conditions depending on the ID. For instance, for a diesel fuel with a low ID, the mixture packets will be richer than stoichiometric before combustion starts, while for the same conditions, a fuel with a longer ignition delay (e.g., a gasoline) will give rise to mixture packets that are much better mixed and much nearer the global mixture strength, which is usually lean in diesel engines at the time of combustion. The details of formation and oxidation of such pollutants are complex and can be found in other text books [e.g., 6.18]. For premixed combustion, where there is no excess oxygen, as in a SI engine, NO_x varies with mixture strength as shown by the solid line in Fig. 6.3, with NO_x peaking at slightly lean mixture strength [6.18]. However, when there is excess oxygen available as in a diesel engine, the rich packets mix with oxygen and burn in a diffusion flame. If all the NO_x were thermal NO_x, and all these mixture packets were to burn out in stoichiometric conditions, there should be very little change in NO_x with mixture strength on the rich side of the curve. However, the mixing and the combustion process in diesel engines are extremely complex [6.18]. Moreover, NO_x levels will be affected by strain rates, interactions between flame fronts, and NO_x destruction [6.19]. Hence, the true variation of NO_x with mixture strength is likely to be qualitatively on the rich side as indicated by the dashed line in Fig. 6.3.

CO and HC (unburned hydrocarbons) are also dominated by mixing and temperature effects [6.20, 6.21]. CO and HC will increase if the local equivalence ratio becomes too lean (overmixing) to permit complete combustion on relevant time scales. In some cases undermixing (e.g., if fuel trapped in crevices enters the cylinder during expansion and does not mix well with oxygen) could lead to locally rich regions that cannot support complete combustion. When high levels of EGR are used, the lack of oxygen can also lead to incomplete combustion. HC can also be produced by wall quenching or bulk gas quenching of the products of incomplete combustion. Lower combustion temperatures also inhibit oxidation of HC and CO after they are formed and increase engine-out concentrations.

6.2 Fuel Effects in Diesel Engines Running on Fuels in the Conventional Diesel Autoignition Range

Conventional, practical diesel fuels have CNs of 40–60 and a boiling range of 160–380 °C, as discussed in Ch. 2. In a modern common-rail direct injection diesel engine with a given fuel at a given speed and load, combustion and emissions depend on engine parameters such as intake pressure, EGR level and the pressure, timing, and strategy of the fuel injection (e.g., the number and phasing of injections). These parameters are optimized to get the desired efficiency and emissions.

Fuel properties such as CN and volatility affect mixing and combustion and, therefore, emissions and efficiency. In addition, chemical composition (e.g., sulfur and aromatic levels) can affect NOx and soot formation. Practical diesel fuels could also contain biodiesels and gas-to-liquid (GTL) components. Biodiesel components have CNs in the 50s and boiling ranges of 300–350 °C. GTL diesel components have CNs greater than 65 and fall within the diesel boiling range, though their final boiling points could be lower. Hochhauser [6.22] reviewed several studies on the effects of diesel fuel properties on emissions from diesel engines. However, many of these properties are correlated in practical diesel fuels, and it is not easy to establish the effect of these fuel properties on emissions independently.

6.2.1 Heat Release and Noise

At low loads, when the amount of diesel fuel needed is low, for a fixed injector and injection pressure the injection duration needs to be small, and the injection event could be finished before combustion starts and PCI combustion is possible. When the mixture packets that autoignite are fuel rich ($\varphi > 1$), the ID will be lower and the pressure pulse will be larger than with leaner mixtures because the autoignition propagation velocity u_a will be larger (see Chap. 5). The mixture packets that autoignite are likely to be fuel rich for fuels of high CN, because the low ID limits the mixing of fuel and air before combustion. Therefore, the peak heat release rate and the temperature will be high. An example of such a heat release rate taken from [6.14] is shown in Fig. 6.4 for a low-load case: 1200 rpm, 4 bar (0.4 MPa) IMEP (indicated mean effective pressure), no EGR for n-heptane which has a CN of around 54. The heat release rate shown is the net heat release rate, which does not account for the heat transferred to the walls. It is calculated from the measured pressure averaged over 250 engine cycles using the first law of thermodynamics, as described in [6.18].

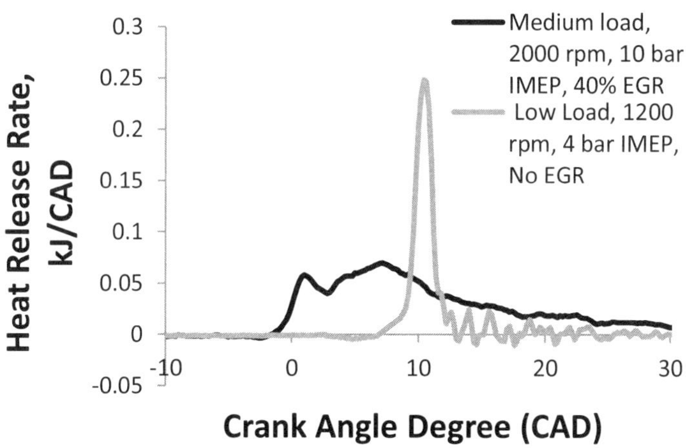

Fig. 6.4. Heat release rates in a single-cylinder engine (compression ratio, CR = 16) at two different loads illustrating premixed and mixing-controlled combustion with fuels of low ID. Results were taken from [6.14]. Low load is with n-heptane, and the medium load is with diesel fuel D1 in Table 6.1. Single-injection, CA50 = ~11 CAD ATDC (crank angle degrees after TDC) for both cases. Maximum heat release rate is much higher at low loads because of autoignition of rich packets as a result of the low ID.

The high combustion temperature will cause high levels of NOx to be formed. Some of the mixture might be rich enough for smoke to be formed, but if there is sufficient oxygen and the temperature is high enough, it will be oxidized inside the cylinder, and engine-out levels will be low. However, if EGR is used to reduce NOx in such conditions, the oxidation of soot will decrease and engine-out soot levels will increase. PCI is possible with conventional diesel fuels only at low loads because the ID is low and the time available for the injection to be completed to get PCI is limited. Section 6.3 will discuss the potential to extend the PCI regime by using fuels with much larger IDs compared to conventional diesel fuels.

At higher loads that require more fuel to be injected, it is not possible to complete the injection event before combustion starts. In such cases combustion occurs in a quasi-steady lifted jet diffusion flame, as described by Dec [6.2, 6.3]. The heat release rate from such combustion has an initial peak that reflects the autoignition of the fuel injected earlier in the injection, which has had more chance to mix with air. Most of the fuel is burned in the predominantly mixing-controlled phase that follows the initial autoignition phase [6.18]. An example of such combustion, again taken from [6.14], is also shown in Fig. 6.4 for fuel D1 in Table 6.1, a European diesel fuel. The heat release rate in the mixing-controlled phase is comparable in magnitude to the turbulent mixing rate [6.18] and is lower when compared to the premixed combustion phase.

The spike in heat release rate associated with premixed autoignition causes a rapid rise in pressure, which is the primary source of noise in diesel vehicles [6.23–6.25]. Measures have to be taken, particularly in LD engines, to mitigate noise at low loads when premixed autoignition predominates. A common way of doing this is to inject a small amount of fuel before TDC (BTDC) [6.26, 6.27] in a pilot injection to be followed by the main injection. Autoignition caused by the pilot injection increases the temperature in the cylinder, causing a reduction in the ID for the main injection. This pushes the combustion from the main injection more toward the mixing-controlled mode, the peak heat release rate is reduced, and average combustion gas temperature is reduced [6.26]. This is illustrated in Figs. 6.5a and 6.5b, showing heat release rates derived from pressure measurements from two cylinders of a European V-6 diesel engine, taken from [6.16] running on fuel D1 in Table 6.1.

Figure 6.5a is the result corresponding to the engine map set by the manufacturer with the pilot injection. In this case the pilot injection started 16 CAD BTDC (crank angle degrees before TDC) and contained about 13% of the total fuel injected. Figure 6.5b shows that the peak heat release rate is much larger

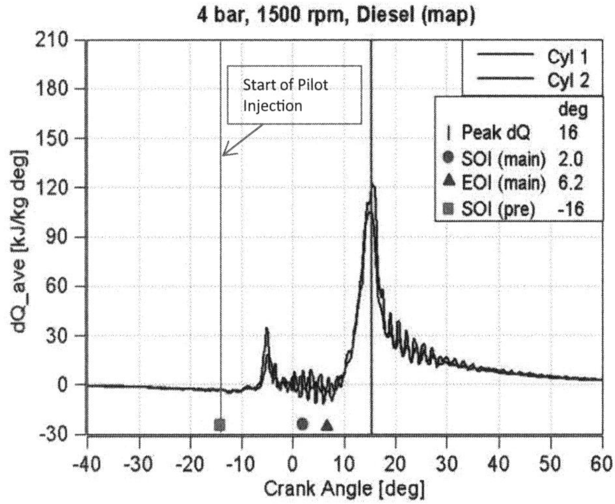

Fig. 6.5a. Heat release rate for cylinders 1 and 2 from a 2.7 L V-6 engine for a European diesel fuel, CN = 54, at the condition set by the manufacturer. Pilot injection was followed by the main injection. 1500 rpm/4 bar (0.4 MPa) IMEP from [6.16]

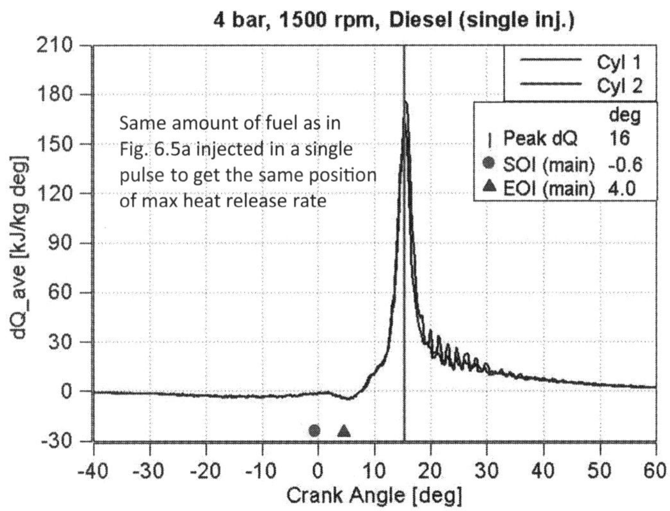

Fig. 6.5b. Heat release rate for cylinders 1 and 2, for diesel fuel, single injection with the position of maximum heat release rate the same as in Fig. 6.5a and same total fuel amount. The peak heat release rate is much larger than in Fig. 6.5a.

when the same amount of fuel as in Fig. 6.5a is injected in a single pulse to get the same crank angle position of the maximum heat release rate. The heat release associated with the pilot injection in Fig. 6.5a occurs before TDC and reduces efficiency, and the promotion of diffusive combustion increases smoke. Thus, it is a common practice in modern LD diesel engines running on diesel fuel to trade reduced noise at low loads for increased smoke and reduced efficiency.

6.2.2 Engine-Operating Parameters, Combustion Phasing, and Emissions in Premixed Compression Ignition with Diesel Fuels

Figure 6.6 shows CA50 plotted against SOI, the CAD (crank angle degree) position of the electric signal which starts the injection, for four fuels in the diesel autoignition range in an engine with a compression ratio (CR) of 16. The data were originally published in [6.12] and [6.14], and the injection rates were such that for all fuels, with SOI at TDC, IMEP was between 4.0 and 4.08 bar (0.4 and 0.408 MPa); the engine speed was 1200 rpm with no EGR; and the injection pressure was 650 bar (65 MPa). IDW was positive in all cases as shown in Fig. 6.7.

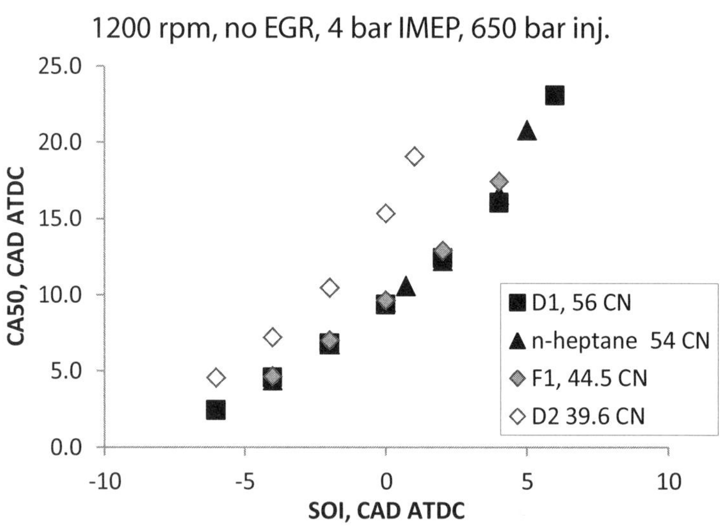

Fig. 6.6. CA50 versus SOI for four fuels in the diesel autoignition range but with different compositions and volatilities. Compression ratio CR = 16, 1200 rpm, no boost, no EGR, injection pressure of 650 bar (65 MPa), and IMEP of 4 bar (0.4 MPa). Original data and engine details are from [6.12] and [6.14]. CA50 increases as start of injection is retarded (i.e., SOI is increased).

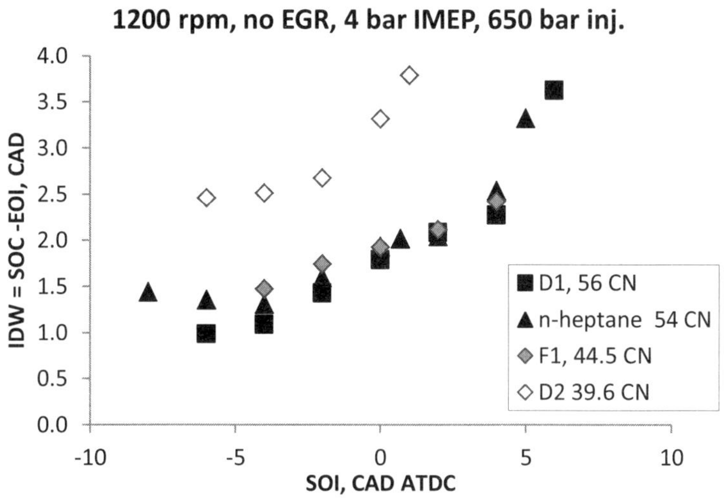

Fig. 6.7. IDW versus SOI corresponding to Fig. 6.6. IDW (SOC − EOI) is positive in all cases—fuel injection is completed before ignition starts.

As SOI increases (i.e., as fuel injection occurs later in the cycle), both CA50 and IDW increase. Combustion phasing parameters such as SOC, CA50, or the position of the peak pressure usually correlate with each other very well, as illustrated in Fig. 6.8, where CA50 has been plotted against CA5 (see also [6.28]). Hence, other measures such as ID and CD are also very well correlated with each other.

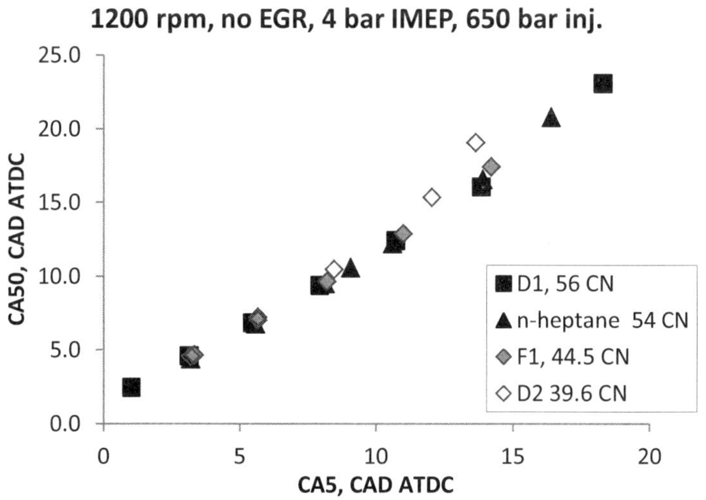

Fig. 6.8. CA50 versus CA5 corresponding to Fig. 6.6. Different combustion phasing parameters are usually very well correlated with each other.

In PCI combustion, the effects of fuel volatility are far less important than the effects of fuel ignitability on combustion phasing [6.14, 6.15, 6.29–6.31]. For instance, there is little difference in Fig. 6.6 between fuels D1, F1, and n-heptane with DCN between 56 and 44 but with very widely varying volatility (Fig. 6.2). Fuel D2 with a DCN of 39.6 showed higher CA50 and IDW for the same SOI, compared to the other three fuels at this operating condition. A fuel does not need to be volatile to enable PCI combustion if it is sufficiently resistant to autoignition. For fuels in the CN range of 40–60, whether there is a noticeable effect of CN on ID and hence on combustion phasing at a given SOI depends very much on the operating and design conditions such as the injection pressure and the compression ratio (CR). For instance, in contrast to Fig. 6.6, there was little difference in CA50 for a given SOI between Swedish MK1 fuel with 54 CN and the 39 CN diesel listed in Table 6.1 with a higher injection pressure of 1300 bar (130 MPa) at similar speed/load conditions in a different engine with a CR of 14 [6.8]. However, when CR was reduced to 11.2 in the same engine at the same conditions (i.e., as autoignition became more difficult), the ignition/combustion delay for the 39 CN fuel was significantly higher than for the Swedish MK1 fuel for a given SOI [6.29]. Similarly, in the presence of EGR, when autoignition becomes more difficult, a fuel of 43 CN had a significantly higher ID compared to a fuel of 53 CN [6.28]. However, if the CN is high and comparable, as with n-heptane and a European diesel, there is little difference in combustion phasing or IDs [6.15, 6.29] in spite of the large differences in volatility and composition.

For a given fuel, in the typical range of injection timings near TDC, advancing the SOI brings the peak heat release rate nearer TDC, increases its magnitude and the maximum pressure rise rate (MPRR). The emissions are also affected by combustion phasing: NOx increases and CO and HC decrease as CA50 decreases [6.8–6.15, 6.28–6.30]. Figure 6.9 illustrates this point for NOx for the case considered in Fig. 6.6.

Combustion phasing and emissions are also greatly affected by EGR: As EGR increases, all else being equal, ID increases and combustion occurs later. However, the major impact of EGR is in reducing combustion temperatures, which causes NOx to decrease. EGR also reduces the oxygen available, which causes CO, HC, and smoke to increase [6.8–6.15, 6.28–6.30]. With longer IDs the mixture packets that burn become leaner at the time of combustion because of the greater time available for mixing. If the mixture packets that burn become less rich on average (i.e., move to the left in Fig. 6.3) for diesel fuel, engine-out NOx will increase in the absence of EGR and soot emissions will decrease. For instance, fuel D2, with a longer ID, shows higher NOx for the same CA50 compared to the other three fuels in Fig. 6.9 in most cases. Other engine parameters, such as injection pressure, that affect the mixing rate also affect emissions. For instance, if the injection pressure is increased, NOx will

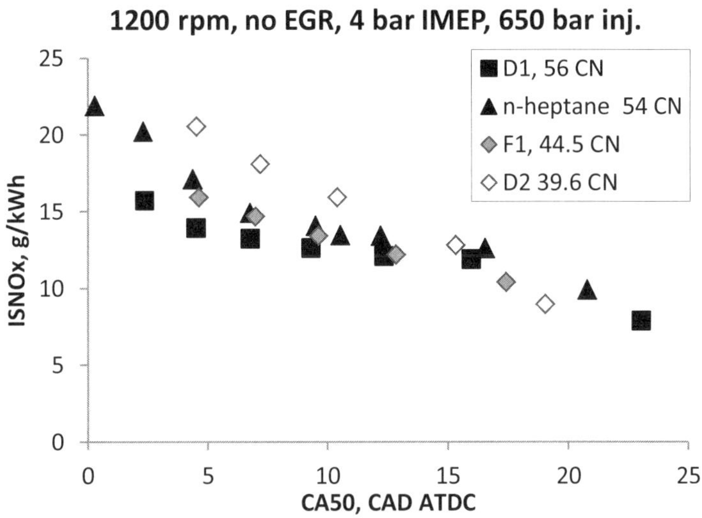

Fig. 6.9. NOx versus CA50 corresponding to Fig. 6.6. NOx decreases with increasing CA50 in PCI combustion. Fuels with similar ignition/combustion delays have similar engine-out NOx.

increase for diesel fuel [6.11, 6.12, 6.32] in the absence of EGR, because the mixture packets that burn move from rich toward stoichiometric mixture strengths (Fig. 6.3) because of better mixing, all else being equal.

In general, if the combustion phasing for a given SOI is comparable for two fuels in PCI combustion, the NOx, CO, and, to a large extent, HC emissions also will also be comparable (see Section 6.3 for further discussion). For instance, in Fig. 6.6, fuels D1, F1, and n-heptane have comparable CA50 for the same SOI and have comparable NOx in Fig. 6.9. Figure 6.10 shows engine-out smoke, which reflects the balance between smoke formation and oxidation, plotted against SOI for the case considered in Fig. 6.6.

Clearly, some smoke is formed with fuel D1 even though IDW is positive, though the engine-out level shown in Fig. 6.10 is quite low. There would be locally fuel-rich mixtures presumably formed by the fuel injected toward the end of the injection event. Fuel D2, the other fuel in the diesel boiling range, has lower smoke because of its higher combustion or ignition delay, which allows it to be better mixed with air at the time of combustion. However, fuel F1 and n-heptane have negligible engine-out smoke. They are both much more volatile than fuels D1 and D2, which would enable better mixing with air and so avoid mixture strengths where smoke could be formed; these two fuels also have no aromatics that are known to increase smoke formation.

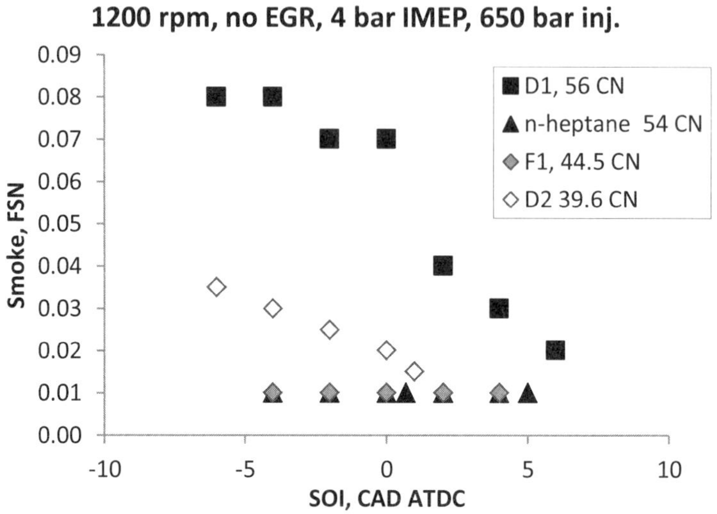

Fig. 6.10. Smoke versus SOI corresponding to Fig. 6.6. Smoke decreases with combustion delay (CD) for fuels in the diesel boiling range. More volatile fuels with no aromatics—n-heptane and F1—have negligible engine-out smoke at this low load.

6.2.3 Higher Loads with Diesel Fuels

Diesel combustion occurs in a quasi-steady lifted jet diffusion flame at higher loads, if the final fuel injection cannot be completed before combustion starts [6.2, 6.3, 6.33, 6.34]. The reaction zone stabilizes at a location downstream of the fuel injector following the initial autoignition. The distance from the injector hole exit to this reaction zone is referred to as flame lift-off length. The oxygen in the surrounding gas is entrained into the fuel jet, mixes with the fuel vapor, and reacts in a fuel-rich premixed flame. The hot products of this rich combustion contain unburned fuel and soot precursors, which react to form soot particles. Much of the soot produced is burned up in the diffusion flame in the periphery of the fuel jet where NOx is also formed. The remainder continues to be oxidized during the expansion stroke. This oxidation will decrease and engine-out soot will increase if the oxygen level and the in-cylinder temperature are reduced (e.g., by the use of EGR) to control NOx. Increasing the lift-off length increases the amount of oxygen available in the fuel-rich premixed flame, and the leaner this flame, the lower the unburned fuel and soot precursors, and the lower the amount of soot that is ultimately produced [6.33–6.35]. Flame lift-off length can be increased by increasing the injection pressure [6.33] or by increasing the ID [6.34].

For the same lift-off length if the engine load is increased by increasing the duration of the fuel injection, more fuel will be burned in the diffusion-controlled mode, the soot level will increase, and the MPRR will decrease.

This is illustrated in Fig. 6.11 where the smoke level is plotted against engine IMEP; the data are taken from [6.9]. A 2-L single-cylinder engine was running at 1200 rpm without EGR on Swedish MK1 fuel from Table 6.1 and a single injection, at a pressure of 1300 bar (130 MPa), started at TDC.

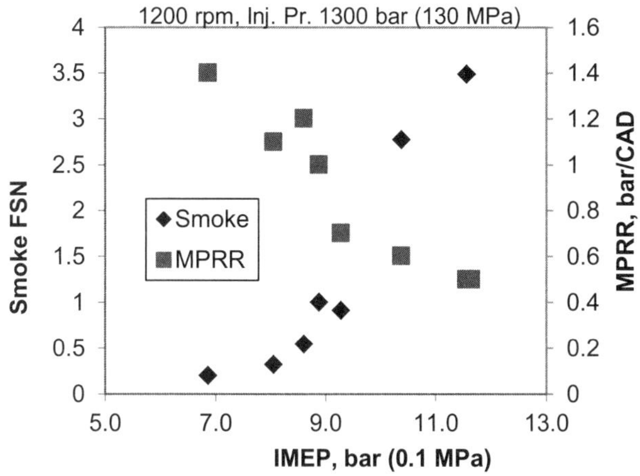

Fig. 6.11. More fuel is burned in the diffusive mode as load is increased by increasing the injection duration. Smoke increases and maximum pressure rise rate (MPRR) decreases. Data from [6.9]. Swedish MK1 fuel, 2-L single-cylinder engine, CR = 14, 1200 rpm, single injection at TDC, injection pressure of 1300 bar (130 MPa), no EGR. CA50 did not differ much for these points.

Even at high loads and high fueling rates, smoke can be reduced for the diesel fuel by injecting it late, after TDC, as illustrated in Fig. 6.12 where the smoke level is plotted against CA50 (filled diamonds) for a fixed nominal fueling rate of 1.2 g/s for the Swedish MK1 fuel.

The corresponding NOx value is low (open diamonds) because the peak temperature in the diffusive mode is low. Figure 6.12 also shows the smoke and NOx results from [6.8] for the same fuel and injection pressure, but at a lower fueling rate of 0.6 g/s where combustion occurred mostly in the premixed autoignition mode. The peak heat release rate and the peak temperature will be higher for the diesel fuel (Fig. 6.4) when combustion occurs in the premixed mode compared to the mixing-controlled mode at higher loads. Hence, at the lower load, MPRR and NOx will be lower, but smoke will be higher, as illustrated in Fig. 6.12, because combustion occurs in the diffusion mode. Fuel composition also affects soot formation and engine-out smoke if soot is formed in the first place because of inadequate mixing (also see Section 6.3.2). For instance, engine-out soot levels decrease with increasing mass fraction of oxygen in the diesel fuel [6.35–6.38], while polyaromatics in the fuel tend to increase soot formation [6.22, 6.39].

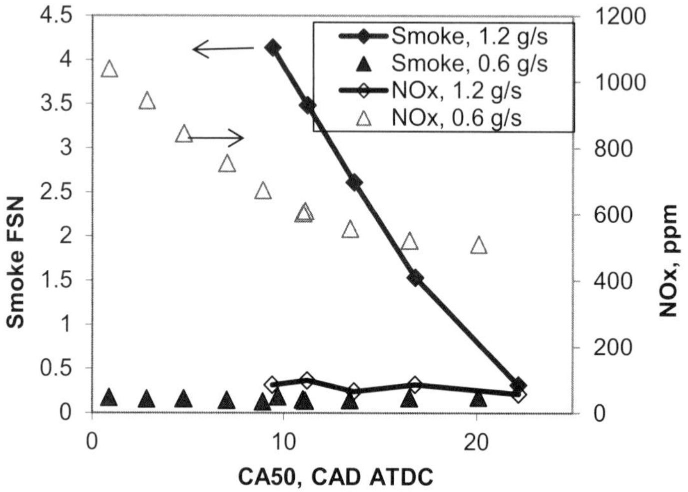

Fig. 6.12. Smoke and NOx versus CA50 for an SOI sweep with Swedish MK1 diesel fuel at two different fueling rates, 0.6 and 1.2 g/s. Combustion was mostly in the premixed autoignition mode for the lower fueling rate and shows higher NOx and lower smoke. Data are from [6.8] and [6.9]. 1200 rpm, 2-L engine, CR = 14, single injection with injection pressure of 1300 bar (130 MPa), T_{in} = 40 °C, CR = 14, no EGR. P_{in} = 1.5 bar (0.15 MPa) absolute for 0.6 g/s and 2.0 bar (0.2 MPa) absolute for 1.2 g/s

6.2.4 Summary on Compression Ignition Combustion with Diesel Fuels

Soot and NOx from diesel engines need to be controlled simultaneously to meet increasingly stringent environmental regulations. NOx can be controlled using EGR, but this will reduce soot oxidation and increase engine-out soot if soot is formed in the first place. Hence, soot formation needs to be minimized or avoided in the first instance. Soot formation cannot be avoided without PCI combustion (Section 6.1), which requires that the final fuel injection is completed well before the SOC to allow all the fuel to be sufficiently mixed with oxygen to avoid local equivalence ratios greater than ~2.

The longer the ID, the easier it is to realize PCI combustion. Conventional diesel fuels, with CNs ranging from 40 to 60, have low IDs and can be run in PCI mode only at low loads. Also, over this CN range, the absolute difference in ID is small and there is little effect of CN on combustion phasing unless other circumstances, such as lower compression ratio (CR) or higher EGR levels, make autoignition more difficult. When PCI combustion is possible with diesel fuels at low loads, premixed autoignition occurs near stoichiometric or rich conditions, and the peak pressure rise rates and NOx are high even though the global mixture strength is lean. Soot that is formed in the rich packets is oxidized and engine-out soot will be low. Practical LD engines have to adopt strategies such as the use of pilot injections that reduce efficiency and increase soot to move the combustion toward a mixing-controlled mode to mitigate noise at low loads. Fuel properties such as higher volatility and lower CNs help make PCI combustion a little easier, but not to the extent required by emissions regulations.

Hochhauser [6.22] has reviewed the extensive literature on the effects of conventional diesel fuel properties on diesel engine-out emissions. However, these fuel effects are small compared to the effects of fuels that are outside the conventional diesel CN range (as will be discussed in Section 6.3). Hochhauser [6.22] also made the point that as engines become more sophisticated and employ emissions aftertreatment systems, fuel effects will be further reduced. Hochhauser's conclusions are summarized in Table 6.2.

Table 6.2 Directional Effects of Conventional Diesel Fuel with CN Range of 40–60 on Engine-Out Emissions as Summarized by Hochhauser [6.22]

	Reduce Density	Reduce PAH*	Increase Cetane	Reduce T90 or T95	Increase Bio Diesel*** Content
HC	Decrease[a]	Increase	Decrease	Increase[b]	Decrease
CO	Decrease[a]	Increase	Decrease	Increase[b]	Decrease
NOx	Decrease[b]	Decrease			Increase
PM**	Decrease	Decrease	Increase[a]	Decrease	Decrease

*Polyaromatic hydrocarbons
**Particulate matter (PM)
***Fatty acid ester
[a] Reliable data only from LD diesel engines [6.22]
[b] Reliable data only from HD engines [6.22]

Modern diesel engines using conventional diesel fuels are forced to use increasingly expensive and complicated technology to achieve low-NOx and low-soot emissions required by modern regulations. For instance, fuel-injection systems employ increasingly higher injection pressures to overcome the low ID of diesel fuels and facilitate the mixing of fuel and air before combustion to reduce soot formation. Aftertreatment systems to reduce NOx and soot will still be required, which increases the cost of diesel engines further.

Several strategies have been proposed to reduce smoke and NOx simultaneously in diesel engines using diesel fuels, and they all rely on minimizing the smoke by promoting premixed combustion. The Toyota UNIBUS system injects some fraction of the fuel very early in the cycle, giving it enough time to mix completely with air. The rest of the fuel is injected near TDC. However, the amount of fuel that can be injected early in the cycle is limited so as to avoid heat release during the compression stroke, and the engine has to switch back to conventional injection timings and conventional diesel or CI combustion at high loads; such a system has been discussed in [6.40, 6.41]. In the Nissan MK (Modulated Kinetics) system, high levels of cooled EGR are used to delay autoignition and reduce temperatures [6.42–6.44]. Hashizume et al. [6.45] introduced the MULDIC (multiple stage diesel combustion) system. In this system part of the fuel is injected very early and undergoes premixed lean autoignition; fuel is injected again well after TDC into the mixture with high temperature. Multiple injection strategies have also been combined with high EGR rates to reduce NOx and smoke [6.46, 6.47]. Hashizume et al. [6.45] also used a fuel in the gasoline autoignition range with a CN of 19 in their experiments. With early injection, diesel fuels will ignite very early in the cycle unless the global mixture strength is very lean [6.47].

Even with multiple injections, the heat release from the first, early, injection will occur during the compression stroke and reduce efficiency [6.45, 6.47] for fuels with high CN. Both smoke and fuel consumption improved when the CN of the fuel used during the first stage was reduced [6.45]. This strategy is comparable to the dual-fuel approach in RCCI (reactivity controlled compression ignition; see Section 6.5). High-swirl, late-fuel (after TDC) injection and low injection duration also contribute to the promotion of mixing prior to ignition in such systems. Such strategies also usually incur penalties in fuel consumption [6.48].

6.3 Premixed Compression Ignition Combustion with Fuels with Long Ignition Delays

Fuels in the gasoline autoignition range (RON > 60, CN < 30) have significantly longer IDs compared to conventional diesel fuels and, as recent experiments [6.8–6.16, 6.49–6.71] have shown, make high-efficiency, low-NOx, and low-smoke operation much easier to achieve.

6.3.1 Low Loads

Figure 6.13 shows CA50 and NOx plotted against SOI for the Swedish MK1 diesel fuel and ULG 95 gasoline (Table 6.1). The data are from [6.8], and the conditions correspond to a fueling rate of 0.6 g/s considered in Fig. 6.12. The IMEP was around 5.7 bar (0.57 MPa) for CA50 between 5 and 7 CAD for both these fuels [6.8] at this condition.

CA50 starts rising sharply for the gasoline upon advancing the injection beyond 30 CAD BTDC, and combustion starts getting very unstable. If the injection timing is earlier than 37 CAD BTDC, the last point on the left shown

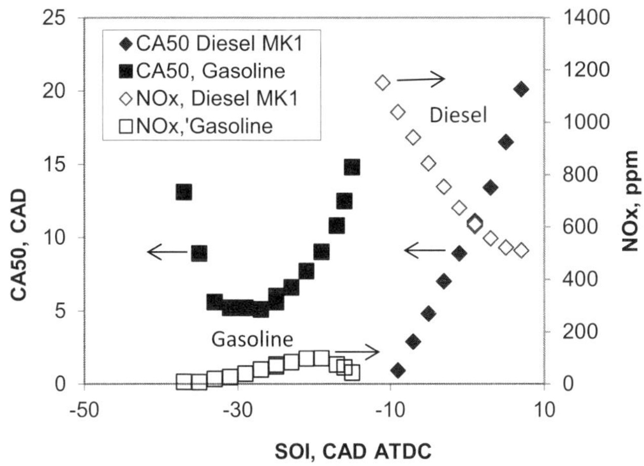

Fig. 6.13. CA50 and NOx versus SOI with Swedish MK1 diesel fuel and ULG 95 gasoline. Data from [6.8]. Operating conditions corresponding to fueling rate of 0.6 g/s in Fig. 6.12. Gasoline has to be injected much earlier than diesel to get the same CA50. Significantly lower NOx with gasoline.

in Fig. 6.13, combustion could not be sustained at all with gasoline. With such early injection, the fuel and air will be more premixed and combustion will approach HCCI combustion. If the fuel and air are fully premixed, the global average mixture strength at this cooperating condition is very lean: $\lambda \sim 3.8$. At such lean conditions, the gasoline used in these tests is too resistant to autoignition to sustain HCCI combustion.

However, with injection starting later than 30 CAD, combustion occurs in a partially premixed mode, and there is obviously some area in the combustion chamber where the mixture is rich enough to ignite, and combustion, once initiated, proceeds through the rest of the mixture. There could be a mixture of autoignition as well as local flame propagation. In this regime, CA50 increases almost linearly with SOI. Thus, there is an injection timing window for the gasoline at this operating condition where the engine will run and combustion phasing can be controlled by the injection timing. The fuel and air are not fully mixed as in HCCI combustion, and the inhomogeneity or stratification in mixture strength that results is absolutely essential to ensure combustion. The diesel fuel cannot be injected very much earlier than shown in Fig. 6.13 either; otherwise, the heat release will occur earlier in the cycle before TDC, and the MPRR will be unacceptably high.

When the ID is large, as for gasoline, there will be more mixing with oxygen before combustion, and the mixture packets that burn will be much nearer the global mixture strength for gasoline compared to the diesel fuel. The mixture packets that burn will be much leaner for high-octane fuels compared to high cetane fuels when the global mixture strength is lean (Fig. 6.3), as it is at low loads. Hence, NOx levels will be significantly lower for gasoline [6.8–6.16] than for diesel, as illustrated in Fig. 6.13. For the same reason, peak heat release rates will be lower for gasoline type fuels at low loads. This is illustrated in Fig. 6.14 for n-heptane and gasoline G4 (84 RON) from Table 6.1. The data are taken from [6.14]: The 0.54-L single-cylinder engine (CR = 16) was running at 1200 rpm without EGR, with a nominal IMEP of 4 bar (0.4 MPa), and CA50 was ~10.5 CAD ATDC for both fuels. The start of the single-fuel injection, with injection pressure of 650 bar (65 MPa), was much earlier for fuel G4 compared to n-heptane: 12 CAD BTDC versus +0.7 CAD ATDC.

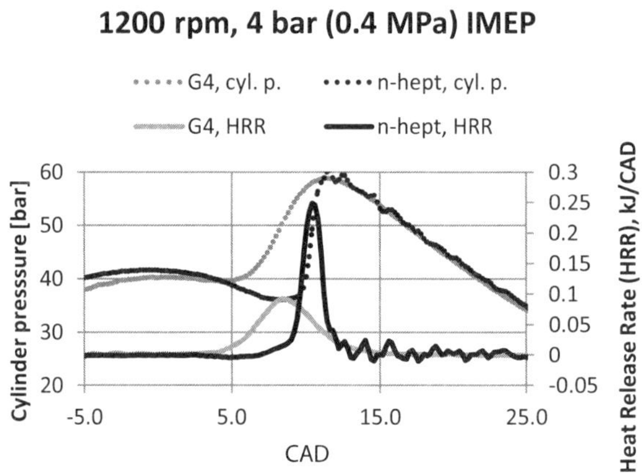

Fig. 6.14. Pressure and heat release rate (HRR) for n-heptane (SOI = +0.7 CAD ATDC) and gasoline G4 from Table 6.1 (SOI = −12 CAD ATDC). Data are from [6.14], 0.54-L single-cylinder engine (CR = 16), single injection with injection pressure of 650 bar (65 MPa), T_{in} = 60 °C, no EGR. P_{in} = 1.1 bar (0.11 MPa) absolute, CA50 = 11 CAD ATDC, nominal IMEP = 4 bar (0.4 MPa).

Hence, the MPRR and noise will be lower for the gasoline compared to the diesel fuel at low load at the same CA50 because the ID is large [6.8–6.16]. This is illustrated in Fig. 6.15 for the condition considered in Fig. 6.14 for the diesel fuel D1 and gasoline G5 (91 RON). For the same reason, HC and CO will be much higher for gasoline compared to diesel. Figure 6.16 shows the corresponding ISHC (indicated specific HC) and ISCO (indicated specific CO) results for the two fuels.

It is common to use pilot injection in diesel engines at low load to reduce noise, but this increases smoke and fuel consumption, as discussed in Section 6.2.1.

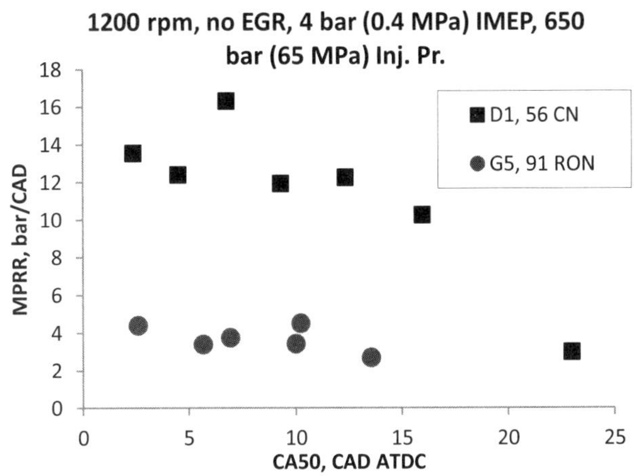

Fig. 6.15. MPRR versus CA50 from an SOI sweep at operating conditions corresponding to Fig. 6.14 for a diesel and a gasoline fuel. Gasoline has a much lower MPRR compared to diesel at low load.

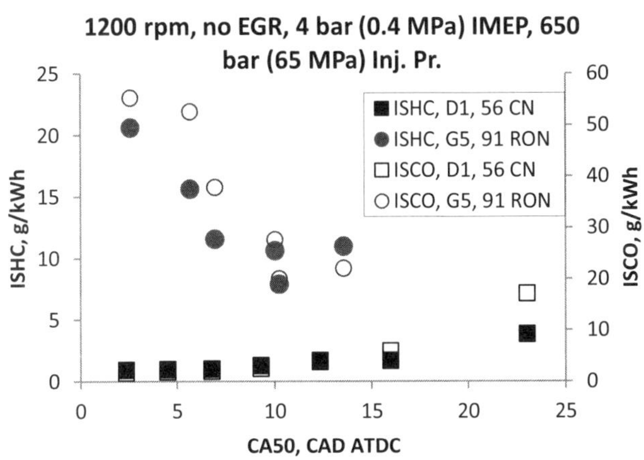

Fig. 6.16. Indicated specific HC (ISHC) and indicated specific CO (ISCO) versus CA50 for diesel and gasoline at operating conditions corresponding to Fig. 6.14. Gasoline has much higher CO and HC compared to diesel.

However, with a gasoline-like fuel this can be avoided, and the overall fuel efficiency of the engine could be improved [6.16].

6.3.2 High Loads

As the load is increased at a given speed by increasing the fueling rate, the global equivalence ratio will increase, and for fuels with high ID, combustion will still be in the PCI mode and the mixture packets that autoignite will be nearer stoichiometric mixture strength than at low loads. In contrast, the diesel fuel will be burning in the diffusive mode because of its low ID. Therefore, the peak temperatures, NOx, and maximum HRRs will be higher for fuels with high ID. EGR can be increased to reduce NOx; ID as well as CD increase for all fuels as EGR is increased. Soot oxidation decreases as EGR increases and engine-out smoke levels go up if soot is formed in the first place. These points are illustrated in the following figures. Figure 6.17 shows combustion delay, CD = CA50 – SOI, plotted against EGR for four very different fuels.

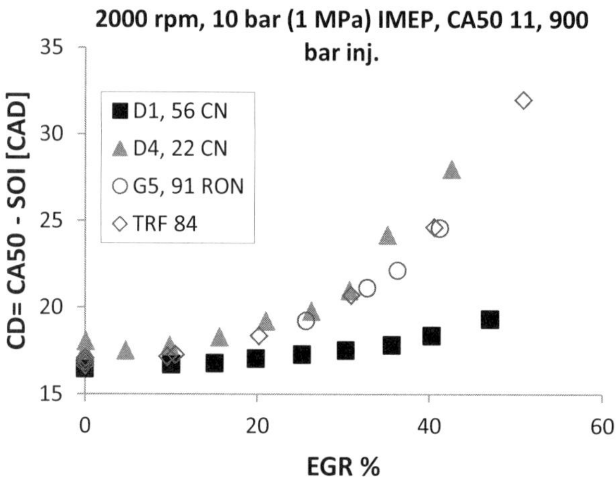

Fig. 6.17. CD versus EGR level for four different fuels. Data are from [6.14] and [6.15]. Fueling rate was fixed to get 10 bar (1 MPa) IMEP with CA50 at 11 CAD ATDC. Then, with fueling rate and CA50 fixed, EGR was varied. 0.54-L single-cylinder engine (CR = 16), injection pressure of 900 bar (90 MPa), T_{in} = 60 °C, P_{in} = 2.0 bar (0.2 MPa) absolute.

The data are from [6.14] and [6.15]. Three of these fuels, gasoline G5 (91 RON), fuel D4, and TRF 84 (a mixture of 65% volume toluene and 35% volume n-heptane) have a high resistance to autoignition compared to the diesel fuel D1. Fuels G5, D4, and TRF 84 have very different composition and volatility. For instance, TRF 84 is a mixture of n-heptane that boils at 98 °C and toluene which boils at 111 °C, while D4 is in the diesel boiling range and has very high (75%) heavy aromatic content. The engine was running nominally at 10 bar (1 MPa) IMEP, 2000 rpm, with a single injection with 900 bar (90 MPa) injection pressure with CA50 fixed at 11 CAD ATDC. With the fueling rate fixed, EGR—defined as the intake CO_2 concentration expressed as a percentage of the exhaust CO_2 concentration—was varied. As the EGR level increased, the injection timing changed to keep CA50 fixed; other details of the experiments are in [6.14] and [6.15]. For all fuels, as EGR is increased, combustion delay increases, but is much lower for the diesel fuel than for the other three fuels. The corresponding IDW = SOC – SOI (see Section 6.1) is shown in Fig. 6.18. It is negative in all cases for the diesel fuel, whereas it is positive for the other three fuels.

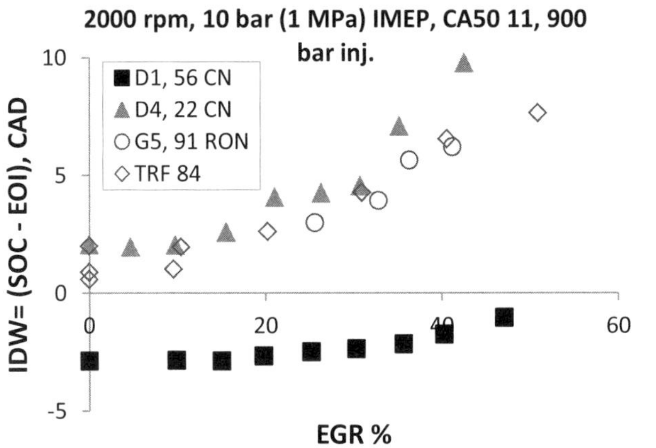

Fig. 6.18. Ignition dwell, IDW = SOC – SOI corresponding to the data in Fig. 6.17. The diesel fuel is burning in diffusive mode (negative IDW), while the other three fuels burn in premixed mode for all conditions.

Thus, the diesel fuel is always burning in the diffusive mode while the other three fuels always burn in the premixed autoignition mode. Figs. 6.19, 6.20, 6.21, and 6.22 show the corresponding indicated specific NOx (ISNOx), smoke level, and indicated specific CO and HC (ISCO, ISHC).

NOx decreases with increasing EGR for all fuels, though NOx is much lower for the diesel fuel at low EGR compared to the other three fuels because of the lower peak temperatures. However, smoke increases with increasing EGR only for the diesel fuel (Fig. 6.20) since smoke formation cannot be avoided because it burns in the diffusive mode. In contrast, for the other three fuels, including fuel D4 with 75% aromatics and very high density (Table 6.1), the smoke level is negligible even at high EGR rates indicating that smoke formation has been avoided in the first place because of PCI combustion. CO is lower (Fig. 6.21) for the diesel fuel as is HC (Fig. 6.22).

Figure 6.23 shows the average heat release for diesel fuel D1 and the 84 RON gasoline, G4, at the same operating conditions as for Fig. 6.17 but with EGR level at 40%.

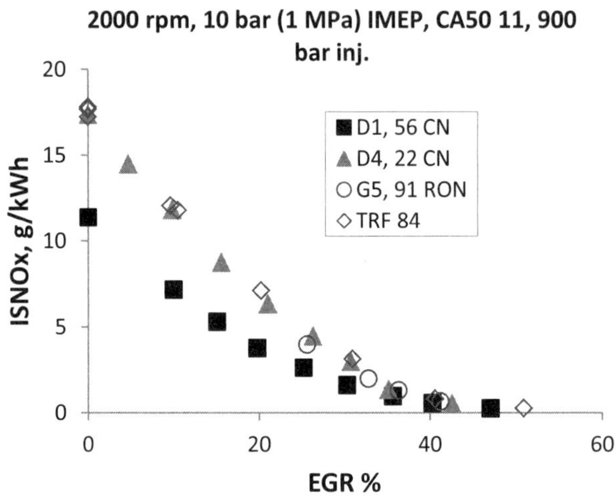

Fig. 6.19. ISNOx corresponding to the data in Fig. 6.17. NOx decreases with increasing EGR for all fuels. The diesel fuel, burning in diffusive mode, has lower NOx than the other three fuels at low EGR because of lower peak temperatures.

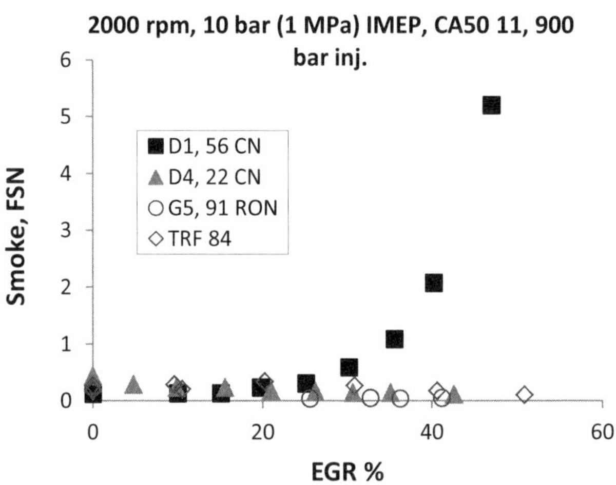

Fig. 6.20. Smoke corresponding to the data in Fig. 6.17. Smoke is always formed for diesel and increases significantly as EGR is increased because its oxidation in the cylinder decreases as temperature and oxygen levels decrease. The other three fuels burn in premixed mode, and thus all, even fuel D4 with 75% aromatics, have negligible smoke at high EGR levels.

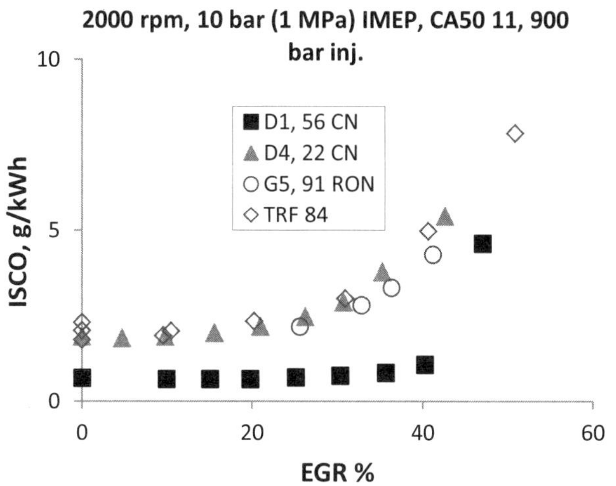

Fig. 6.21. Indicated specific CO (ISCO) corresponding to the data in Fig. 6.17. CO increases with increasing EGR for all fuels as the temperature and oxygen level decreases. CO is lower for diesel fuel at this condition

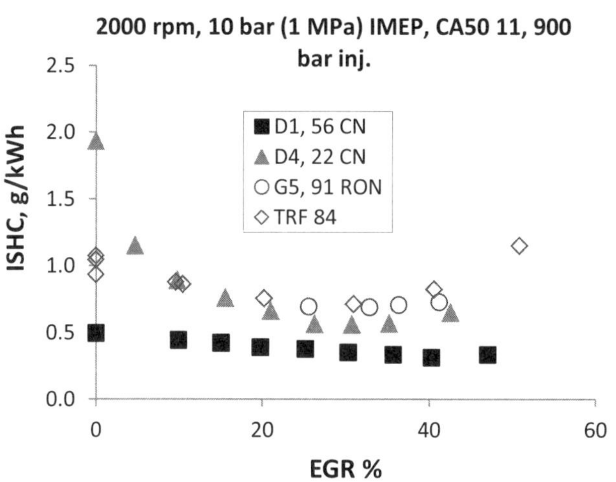

Fig. 6.22. Indicated specific HC (ISHC) corresponding to the data in Fig. 6.17. HC is lower for diesel fuel at this condition

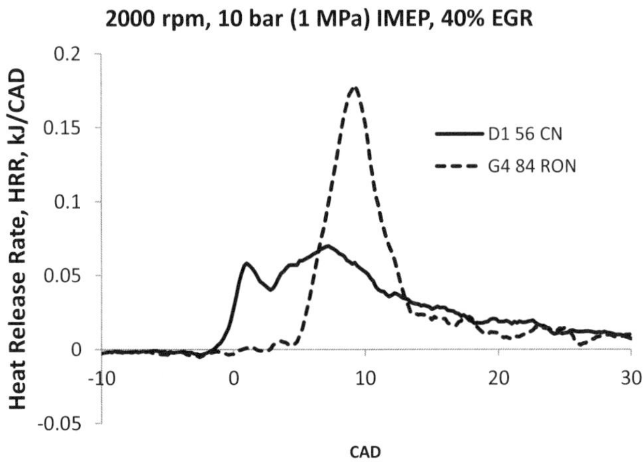

Fig. 6.23. In contrast to Fig. 6.14, the diesel fuel, which burns in diffusive mode, has lower maximum HRR compared to the gasoline, which still burns in premixed mode because of the longer ID at this condition.

The peak HRR is higher for the gasoline, which burns in premixed mode, compared to the diesel fuel, which burns in the mixing-controlled mode. This is in contrast to the low-load condition where the diesel fuel has a higher peak HRR, as discussed in Section 6.3.1. As a result, the MPRR is lower for the diesel fuel (Fig. 6.24).

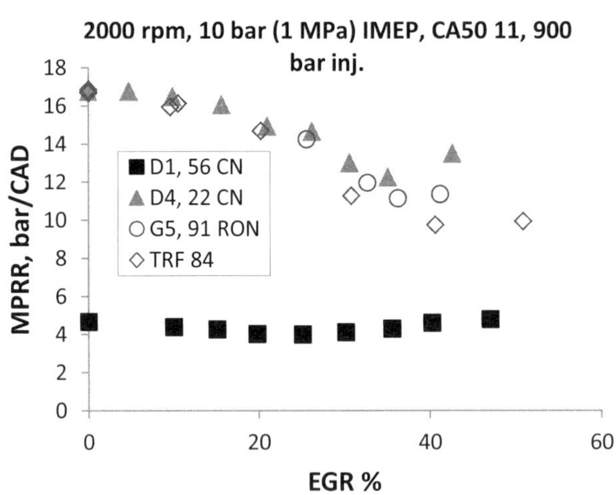

Fig. 6.24. MPRR corresponding to the data in Fig. 6.17. MPRR is lower for the diesel fuel, which burns in mixing-controlled mode at this condition.

Figure 6.25 shows the corresponding indicated specific fuel consumption (ISFC).

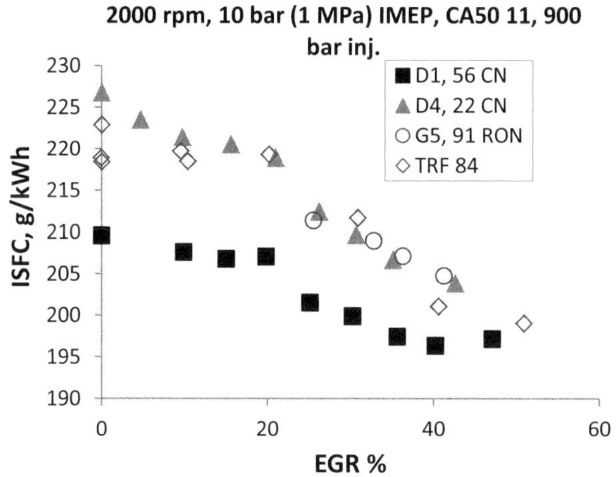

Fig. 6.25. ISFC corresponding to the data in Fig. 6.17. ISFC decreases with increasing EGR when CA50 is fixed and is lower for the diesel fuel compared to the other three fuels.

In these experiments, fuel flow rate was calculated from measured air consumption rate and the estimated equivalence ratio based on exhaust emissions rather than by direct measurement. When CA50 is kept constant, ISFC decreases with increasing EGR [6.9–6.12, 6.14], a trend also noted in [6.52] and attributed to reduced heat losses because of lower combustion temperatures with increasing EGR. ISFC is lower for the diesel fuel compared to the other three fuels at this condition partly because of the higher HC and CO emissions for these fuels. However, the difference between the fuels decreases as the EGR level increases. Moreover, ISFC, HC, and CO can be reduced for gasoline-like fuels with high ID without increasing the smoke level significantly by reducing the injection pressure (see Section 6.3.4).

6.3.3 Effects of Gasoline Fuel Properties on Premixed Compression Ignition Combustion

In PCI combustion, if two fuels have the same combustion phasing (e.g., CA50) at the same injection timing, they have comparable NOx, smoke, CO, ISFC, and MPRR regardless of any differences in volatility or composition [6.12–6.15]. There might be some differences between such matching fuels in terms of hydrocarbons, but these differences are small compared to fuel pairs with very different combustion phasing. It is likely that CO and NOx are dominated by processes in the bulk gas, while hydrocarbon emissions are also affected by the processes in the quench and crevice layers and might be more affected by volatility and compositional differences.

For instance, in Fig. 6.17, for a given EGR level, CD = CA50 – SOI, and hence, since CA50 is fixed, SOI is the same for the three fuels, D4, G5, and TRF 84, which have very different volatility characteristics as well as composition. These fuels show comparable NOx, CO, HC, MPRR, smoke, and ISFC for the same EGR level (Figs. 6.18–6.22). Such results are also in line with observations in [6.30, 6.31] that, in low-NOx, low-smoke combustion in CI engines, fuel autoignition quality—which determines combustion phasing—is far more important than its volatility and composition. In general, if two fuels have the same ID or CD for the same SOI, they will have comparable emissions and efficiency in PCI combustion.

The octane index (OI) approach discussed in Chapter 4 is also found to be applicable to PCI combustion [6.13, 6.15]. The true autoignition resistance of a gasoline is described by its octane index with this equation: $OI = \text{RON} - KS$, where $S = (\text{RON} - \text{MON})$ is the sensitivity, RON and MON are the research and motor octane numbers, and K is a constant depending on the pressure and temperature history of the fuel/air mixture. Parameters such as CD or ID, which depend on the autoignition resistance of the fuel at a given operating condition, correlate with OI. Nonparaffinic fuels have $S > 0$. As discussed in Chapter 4, if the pressure of the mixture at a given temperature is higher than in the RON test, the value of K is negative, and nonparaffinic fuels become relatively more resistant to autoignition compared to paraffinic fuels (i.e., their OI > RON), and they behave like a primary reference fuel whose octane number is greater than the RON of the fuel.

In SI and HCCI combustion the mixture strength remains unchanged, but the pressure and temperature increase until autoignition occurs. In PCI combustion, after the fuel is injected into a high-pressure and high-temperature environment, the pressure and temperature do not change much, but the mixture strength changes in both time and space until ignition occurs. Since autoignition reactions occur essentially at much higher pressures, we should expect nonparaffinic fuels to be relatively more resistant to autoignition (see the discussion in Section 4.1.2), and K to be negative. This is indeed the case, as illustrated in Figs. 6.26–6.28 with the data from [6.13]. In the tests considered here, different fuels including full-boiling-range gasolines and primary and toluene reference fuels were run at the operating conditions considered in Fig. 6.14. These fuels have very different compositions and hence different RON and sensitivity values. Only those fuels in the gasoline autoignition (RON > 60) and volatility range are considered for analysis. Fuel D4 might behave like a gasoline, as discussed in the previous section, but its RON and MON cannot be determined in the standard CFR tests because of its low volatility, and so it is not considered.

Figures 6.26, 6.27, and 6.28 show the combustion delay (CD) with CA50 fixed at 11 CAD ATDC, plotted against RON, MON, and OI. The *K* value was estimated by linear regression (as discussed in Section 4.2.3) with *z* in Eq. 4.4 replaced by CD. CD has a poor correlation with RON or MON, but OI with $K = -2.95$ correlates with CD very well.

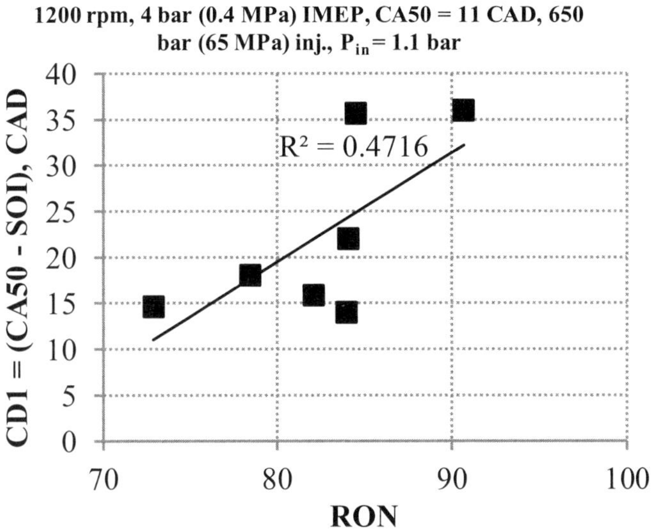

Fig. 6.26. Combustion delay (CD) plotted against RON. Experimental conditions were as in Fig. 6.14 and results were taken from [6.13].

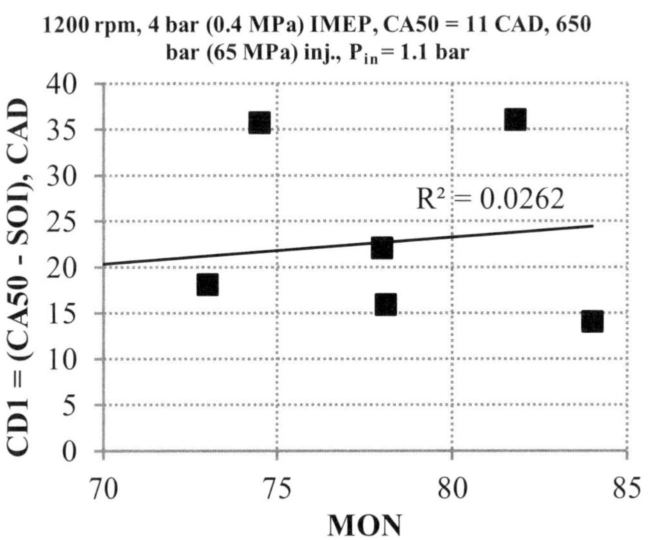

Fig. 6.27. CD from Fig. 6.25 plotted against MON.

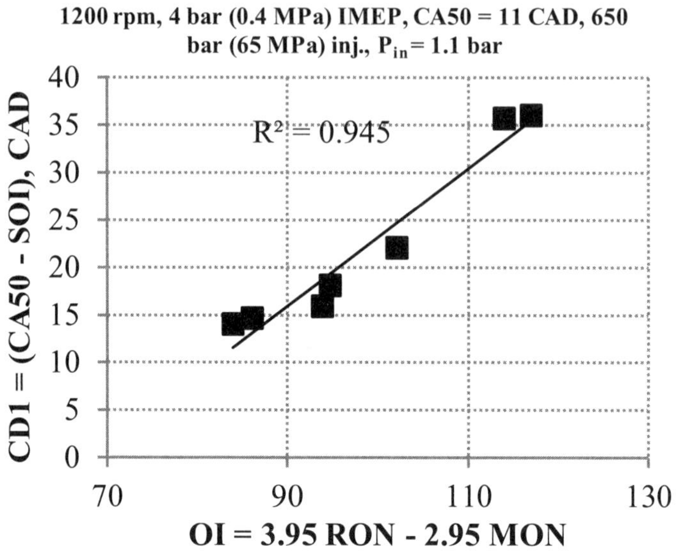

Fig. 6.28. CD from Fig. 6.25 plotted against octane index, $OI = (1 - K)\text{RON} + K\text{MON}$, with $K = -2.95$.

As the OI of the fuel is increased, say, by increasing the RON at a given operating condition, the ID increases. At any operating condition, there is a limiting ignition delay (IDL). If the ID is higher than IDL, ignition will fail. As the actual ID is reduced below IDL, ignition will occur earlier in the cycle. IDL itself depends on the engine-operating conditions. Autoignition is difficult at low pressures and temperatures, at low loads when the global mixture strength is lean, and at high EGR levels or at higher speeds when the time available for ignition is low. IDL will be low when autoignition is difficult. IDL will also be reduced when the CR of the engine is reduced. Fuels with high OI (high RON) will be difficult if not impossible to run at conditions when IDL is low. At the same time, ID should be high enough to enable PCI combustion. For an engine with a CR of ~16, Hildingsson et al. [6.11] suggested that the optimum fuel for PCI combustion would have a RON between 75 and 85. In contrast, in an engine with a CR of 12, the fuel had to have much lower RON of around 60 to enable PCI to be run over the chosen operating points that covered the FTP75 city driving cycle [6.60]. Manente et al. [6.51] suggested that the optimum RON for PCI engines is "in the range of 70"; CR was either 14.3 or 17 in their experiments. Higher-octane fuels make low-load, high-EGR operation more difficult while lower-octane fuels reduce the load at which PCI operation is possible.

In a multicylinder engine carrying a single gasoline-like fuel, variable valve actuation can be used to change in-cylinder conditions to enable the engine to operate over a wide operating range in PCI mode [6.54, 6.55]. At low loads, exhaust rebreathing by opening the exhaust valve during the intake increases the in-cylinder temperature and makes autoignition easier while at high loads, closing the intake valve late reduces the effective CR, and thus reduces the pressure and temperature in the cylinder, and increases the ID [6.54, 6.55].

Though volatility is not as important as autoignition resistance, it still plays a role since it will affect mixing. There is some evidence that the wider volatility range of the gasoline–diesel blends enables more stratification and is beneficial. For instance, the gasoline fuel G5 in Table 6.1 could not be run with an EGR level greater than 37% at 3000 rpm, 10 bar (1 MPa) nominal IMEP [6.12, 6.14]. In contrast, a mixture of 10% (volume) of European diesel and 90% (volume) of 95 RON gasoline, GD10, which had very similar RON and MON as G5 and—hence could be expected to behave very similarly to G5—could be run at higher EGR levels and even in the conditions at 4000 rpm, 10 bar (1 MPa) nominal IMEP at high EGR levels [6.62]. Further work is needed to clarify this point.

6.3.4 Effects of Injection Pressure, Injection Strategies, and Injector Design on Gasoline Premixed Compression Ignition

As discussed in Section 6.3.2, the maximum HRR, and hence the MPRR (which affects engine noise), is high at high engine loads. Double [6.9] and triple [6.10] injection strategies can greatly help in reducing the peak pressure rise rate at high loads while still meeting the requirements for low NOx, low smoke, and acceptable efficiency. Indeed, most recent studies in this field employ multiple injection strategies in gasoline PCI [6.49–6.52, 6.54, 6.55] at medium and high loads. Significant amounts of gasoline-like fuel can be injected fairly early in the cycle without causing any heat release during the compression stroke because of the high resistance of the fuel to autoignition. The final fuel injection nearer TDC then triggers the combustion, presumably by introducing the required inhomogeneity in the mixture. Thus, multiple injection of a gasoline-like fuel allows the spatial stratification of the mixture in the cylinder to be better controlled so that the heat release can be more spread out in time, and the peak HRR is reduced. At any speed/load point, such a multiple injection strategy needs to be optimized in terms of the number of injections and the timing and duration of each of these injections to meet the required emissions and efficiency targets. Moreover, the optimum strategy will also depend on other operating parameters such as EGR rate, injection pressure, swirl ratio, and boost pressure, which affect mixing of fuel and oxygen in the cylinder [6.54, 6.55].

At low loads, when the global mixture strength is very lean, and when the ID is large, the mixture packets that burn are lean, and CO and HC are high (as discussed in Section 6.3.1). Reducing the injection pressure has a significant benefit in improving the low-load PCI operation of gasoline-like fuels presumably by reducing the chances of overmixing and overleaning. This is illustrated in Figs. 6.29–6.31, where fuel G4, the 84 RON gasoline is compared with the European diesel D1 at 2000 rpm, 2 bar (0.2 MPa) absolute inlet pressure at different injection pressures.

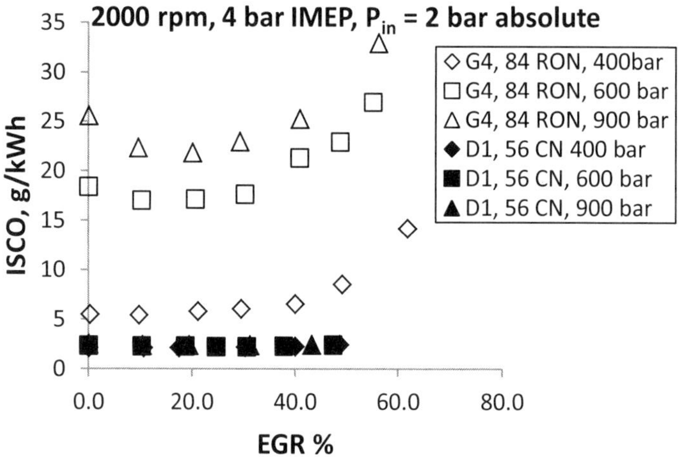

Fig. 6.29. ISCO versus EGR level for gasoline G4 (84 RON) and diesel D1 at different injection pressures (data are from [6.11]). Significant reduction in CO for the gasoline as injection pressure is reduced.

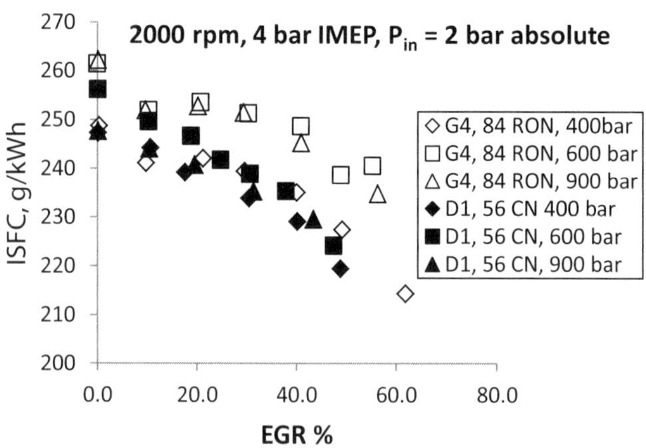

Fig. 6.30. ISFC versus EGR level for gasoline G4 (84 RON) and diesel D1 corresponding to Fig. 6.29. ISFC also decreases for gasoline as injection pressure is reduced.

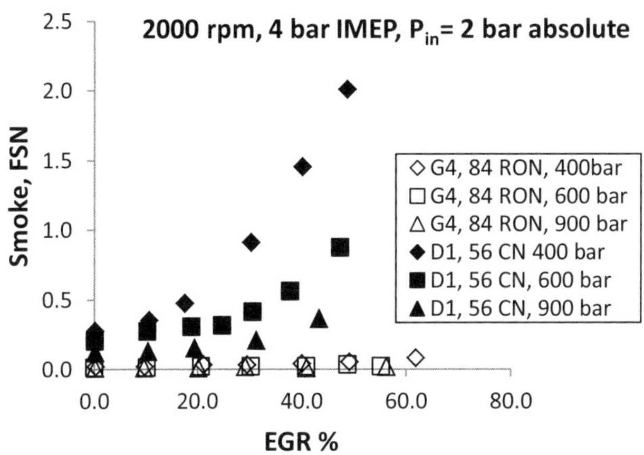

Fig. 6.31. Smoke versus EGR level for gasoline G4 (84 RON) and diesel D1 corresponding to Fig. 6.29. As injection pressure is reduced, smoke increases significantly for diesel but still remains very low for the gasoline.

In these experiments, fully described in [6.11], at each injection pressure the fueling rate was fixed to yield 4 bar (0.4 MPa) IMEP without EGR with CA50 at 11 CAD ATDC. Then EGR was varied with the fueling rate and CA50 fixed. As the injection pressure decreases and the injection pulse width increases, IDW decreases. Reducing the injection pressure from 900 to 400 bar (90 to 40 MPa) significantly reduces CO emissions (Fig. 6.29) and ISFC (Fig. 6.30) for the gasoline without really affecting the smoke emissions (Fig. 6.31). ISNOx is reduced by increasing EGR for both fuels and does not depend on injection pressure as shown in Fig. 6.32.

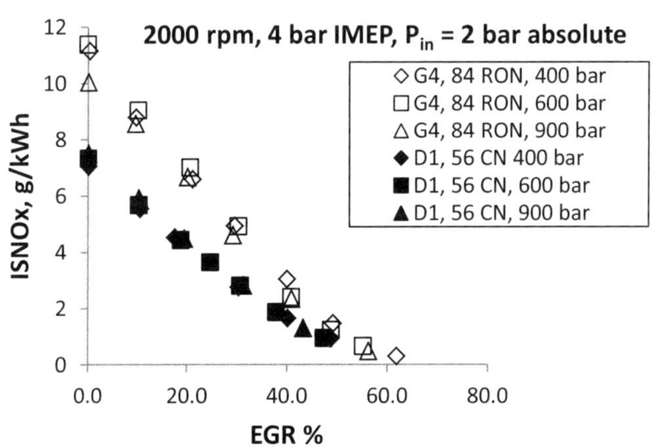

Fig. 6.32. ISNOx versus EGR level for gasoline G4 (84 RON) and diesel D1 corresponding to Fig. 6.29. As EGR is increased, NOx is reduced for both fuels at all injection pressures.

Similar results are shown for fuel G5, the 91 RON gasoline in [6.12]. However, for the diesel fuel, as the injection pressure is decreased, smoke increases significantly (Fig. 6.31). Even at higher loads, PCI combustion can be obtained with gasoline using low injection pressures as demonstrated in [6.54, 6.55], where the injection pressure used was below 500 bar (50 MPa) and in [6.63] where the maximum injection pressure used was 400 bar (40 MPa).

Almost all the work in this field has so far been done using modern diesel injectors with relatively small nozzle holes. Given that the mixing is dominated by the large ID available and overmixing is more of a problem at low loads, larger nozzle holes coupled with lower injection pressures might be better suited for gasoline PCI. Even at higher loads, a larger nozzle-hole size would reduce the injection duration at a given injection pressure and facilitate PCI combustion. Indeed, preliminary tests in a single-cylinder engine [6.61, 6.62] showed that the operating limits could be extended by increasing the nozzle-hole size and using lower injection pressures. In [6.60] PCI combustion was demonstrated using a gasoline DISI injection system.

Clearly, a significant amount of development work is needed to optimize the injector design and injection strategies for gasoline PCI.

6.3.5 Dual-Fuel Operation or Reactivity Controlled Compression Ignition

The required autoignition quality of the fuel changes as engine-operating conditions change. At low loads, when the overall mixture strength is very lean or when high levels of EGR are needed to control NOx, autoignition is difficult and the IDL is low. Under such conditions, a fuel that autoignites easily (i.e., a fuel with low OI) will be required by the engine. On the other hand, at higher loads, when the mixture packets, after mixing, will be nearer stoichiometric conditions, autoignition will be easier, and a fuel with a low OI will lead to a short ID and PCI combustion will be more difficult. Under such conditions, a fuel with higher OI will be needed to get PCI. One approach that overcomes this difficulty is to carry two fuels of different autoignition characteristics and change the autoignition quality of the fuel onboard as required. Reitz and coworkers [6.65–6.71] have termed this dual-fuel strategy RCCI.

In this approach, a fuel with high resistance to autoignition such as commercially available gasoline or mixtures of gasoline and ethanol [6.67, 6.70] or natural gas [6.71, 6.72] is injected in the port, and ignition is triggered by the direct injection of a fuel of high reactivity such as commercially available diesel fuel near TDC. The ratio of the two fuels used is changed depending on the engine-operating condition (e.g., relatively more diesel fuel is used at low loads). However, over a normal operating range the amount of diesel fuel used is around 10% of the total fuel used. A variation on this theme is to carry one gasoline-like fuel onboard but use varying amounts of a diesel ignition improver to make the fuel more reactive and inject this reactive fuel directly into the cylinder to trigger ignition [6.65, 6.70]. As summarized in [6.69], RCCI combustion is capable of operating over a wide range of engine loads with near zero levels of NOx and soot, acceptable pressure rise rate and noise, and very high indicated efficiency.

6.3.6 Full Load Operation on Gasoline Compression Ignition

In LD passenger car diesel engines, NOx and soot are controlled over an operating cycle—up to a limiting load at any given speed where PCI combustion is possible. This limiting load can be significantly increased using gasoline-like fuels so the average speed of the duty cycle can be reduced. This affords the possibility of downspeeding or downsizing the engine. Beyond the limiting load, the engine has to be run in a mixing-controlled mode by injecting sufficient fuel early in the cycle to produce heat release near TDC and then injecting more fuel into the burning gases. However, emissions regulations for HD engines, particularly in the United States, require that NOx and soot are controlled even at full load at all operating speeds and these speeds are lower than for passenger car engines. Low-NOx and low-soot operation meeting these stringent requirements has been demonstrated using gasoline-like fuels in heavy-duty single-cylinder and multicylinder engines [6.51].

6.3.7 Fuel Efficiency with Gasoline Compression Ignition

Very high efficiencies have been reported with CI combustion using gasoline-like fuels. Indicated fuel efficiencies of up to 56% have been measured with single fuel [6.51, 6.52] as well as dual-fuel operation [6.65–6.70] in HD engines running at medium to full load. In LD engines, compared to running the engine on gasoline in SI mode, fuel consumption could be reduced by about 20% over an operating cycle when the engine is run in CI mode on the same gasoline [6.54, 6.55] or a much lower octane fuel [6.60]. There could be further fuel efficiency gains, compared to running the LD engine on diesel fuel, because the engine could be downsized because NOx and soot can be controlled at higher loads at a given speed using gasoline. Also, in LD engines, use of pilot injection at low load to control noise could be avoided when gasoline is used (Section 6.3.1), leading to further improvements in efficiency. Thus a gasoline CI engine could be at least as efficient as a diesel engine while meeting the emissions and other targets at lower cost, compared to running the same engine on conventional diesel fuel.

6.3.8 Modeling Gasoline Premixed Compression Ignition

There are many engine-operating parameters on a given engine, such as speed, injection pressures, injection strategies, EGR levels, valve timings, intake pressure, and temperature, that can be varied to meet required targets for efficiency, emissions, and pressure rise rates. Moreover, design parameters such as engine CR and injector designs need to be optimized. It would be prohibitively expensive to find the optimum combination of these many parameters by a purely empirical, experimental approach. Reliable models that can guide experiments and optimization would be valuable in the development of this approach. Some examples of such modeling efforts are reported in [6.56, 6.69, 6.73–6.75].

Mixing of fuel and air before combustion determines the combustion and emissions in PCI operation. Hence, the models need to predict the spatial and temporal development of the mixture in the cylinder and combine this with an appropriate kinetic model that predicts autoignition in the cylinder. The kinetic model needs to be small enough so that it can be merged with a mixing model. If the chemical kinetics of a gasoline are to be modeled in PCI combustion, a mixture of iso-octane, n-heptane, and toluene that has the same RON and MON and whose composition could be determined as described in [6.76] would be a very good surrogate for the gasoline [6.15]. It would have the same OI as the gasoline and thus the same combustion phasing with a given injection strategy and, as discussed in Section 6.3.3, would have comparable performance and emissions in PCI combustion.

6.4 Summary and Concluding Remarks

CI engines are more efficient than SI engines. Currently, the only practical CI engine is the diesel engine running on conventional diesel fuels, which have high CNs (> 40) and are very prone to autoignition. Thus, at most operating conditions combustion starts very soon after the fuel is injected, before it has had enough time to mix with oxygen in the cylinder. Compared to SI engines, HC and CO are much lower in diesel engines, but soot and NOx levels are higher.

The global mixture strength is leaner than stoichiometric at all operating conditions in diesel engines, and reduction of NOx in the exhaust using a catalyst is particularly difficult in the presence of oxygen. Equally important, soot formation cannot be avoided, because locally rich mixtures, with equivalence ratio greater than 2, cannot be avoided at most operating conditions in diesel engines running on conventional diesel fuels. The soot formed is also oxidized in the cylinder, and the engine-out soot level reflects the balance between the formation and oxidation. NOx can be reduced in the cylinder by reducing combustion temperatures through the use of EGR, but this reduces the oxidation of soot and increases engine-out soot. Hence, in diesel engines using conventional diesel fuels, it is very difficult to control NOx and soot emissions simultaneously, and the emissions regulations on these pollutants are becoming increasingly stringent throughout the world. If soot formation is avoided in the first place, NOx can be controlled with the use of EGR.

If the last fuel injection is completed sufficiently in advance of the SOC, soot formation can be avoided, and this can be termed PCI combustion. PCI combustion is only possible at very low loads with conventional diesel fuels. Modern diesel engines employ technology such as high injection pressures to expand the operating region where PCI combustion is possible, but this increases the cost significantly and increases parasitic losses, which impact the overall efficiency to some extent. There is also a trend to reduce CRs in diesel engines in order to increase IDs and make PCI combustion easier. Further aftertreatment of the exhaust will usually be required to meet stringent emissions regulations on soot and NOx, increasing the cost further and reducing fuel efficiency to some extent (e.g., during the regeneration of the diesel particulate filter, or DPF).

PCI combustion is much easier with fuels in the gasoline autoignition range (RON > 60) because the higher ID allows more time for mixing of the fuel and oxygen before combustion starts. The approach described in Section 6.3 can be termed GCI (gasoline compression ignition). Compared to an advanced diesel engine, GCI is likely to require much lower injection pressures and a simpler aftertreatment system since the focus shifts to oxidizing HC and CO in an oxygen-rich environment rather than controlling NOx and soot simultaneously. At a given speed, NOx and soot could be controlled at a much larger load in GCI compared to CI combustion using conventional diesel fuels. This could enable downspeeding and downsizing in LD diesel engines that have to meet NOx and soot emissions targets over an emissions test operating cycle and lead to fuel efficiency gains.

Corrective action such as pilot injection is employed in LD diesel engines at low loads to mitigate noise. This could be avoided in GCI and could also lead to fuel efficiency gains. Thus, from a diesel engine perspective, the cost could be reduced without compromising on emissions requirements by running the diesel engine on gasoline-like fuels with possible improvements in efficiency.

From an SI engine perspective, GCI can be considered a way of improving the efficiency of an engine carrying a gasoline-like fuel. A related approach to improve efficiency and achieve low NOx and low soot in an engine fueled by gasoline is to run the engine in HCCI, that is, fully premixed CI mode where possible, but run in SI mode where it is not possible (e.g., at startup and warm-up and at higher loads). GM and Mercedes-Benz (DiesOtto) have announced such initiatives [6.77]. The CR of an HCCI/SI engine is limited because of knock in the SI mode at high load, whereas the GCI engine will run in CI mode at high load and can have a CR comparable to that of a diesel engine. There is in-cycle control of combustion phasing in GCI, via the injection timing of the final injection event, making overall engine control easier compared to HCCI. Other more common approaches toward improving the efficiency of the SI engine move the engine design and operation toward the diesel engine (e.g., turbocharging allied with down-sizing) to minimize throttling losses, running the engine lean and direct injection. These approaches increase the cost of the SI engine but are also limited by knock and flammability limits.

Moreover, GCI engines could use poor-quality gasoline, with RON between 70 and 85 and relatively easy requirements on volatility. Such a fuel would require much less processing compared to today's gasolines. Indeed, if diesel engines continue to use conventional diesel fuel, there is likely to be a very significantly larger increase in demand in the future for diesel fuel compared to gasoline, as discussed in Chapter 1. Refineries will have to accommodate this demand imbalance even as they process more crude to meet the overall increase in fuel demand. In those circumstances there will be greater availability of light fractions such as straight-run gasoline, the mixture of components from initial distillation of the crude oil, which usually have octane numbers around 70. Such fractions can be used in GCI engines with little further upgrading in terms or either cetane or octane.

Table 6.3 compares some of the key general features for GCI engines, advanced diesel engines using conventional diesel fuels, HCCI/SI engines, and SI engines.

So the advantages of the GCI concept are the following:

1. The engine will be at least as efficient and clean as current diesel engines, but will be cheaper (e.g., lower injection pressures, aftertreatment requirement shifted from NOx control to HC and CO control).
2. The optimum fuel will be less processed and hence simpler to make compared to current gasoline or diesel fuels.
3. It provides a path to mitigate the global demand imbalance between heavier and lighter fuels than is otherwise projected. The alternative is investment in refineries to make the required diesel fuel while the diesel engine continues to be expensive in order to meet stringent NOx/soot requirements. Another alternative would be to bring more natural gas into the transport sector in the form of compressed natural gas (CNG) or GTL to replace conventional diesel fuel.

Table 6.3 Comparison of Some General Characteristics between Different Combustion Systems for Internal Combustion Engines

	GCI	Advanced Diesel Running on Diesel Fuel	HCCI/SI Running on Gasoline	SI
Operating mode	CI at most operating conditions; SI at start and low load a possibility	CI	HCCI at low load, SI at startup and warming and high load	SI
Load control via	Fueling rate	Fueling rate	Fueling rate in HCCI, throttle in SI	Throttle
Fuel-injection pressure	Moderate	High	Low	Low
Efficiency	High	High	Medium	Low
Aftertreatment focus	HC and CO	Soot and NOx	HC, CO, and NOx from SI	HC, CO, NOx; three-way catalyst
Overall Cost	Low/moderate	High	Moderate	Low but increasing in the quest for higher efficiency
Fuel	Market gasoline or gasoline/diesel mixtures to start with but in the future low octane—RON between 75 and 85—fuel with no large constraints on volatility	High-cetane diesel	High-octane (RON) gasoline	High-octane (RON) gasoline

However, a significant amount of development work is needed before a GCI engine can power a practical vehicle. Some of the more important problems to be solved include starting the engine at cold conditions, ensuring smooth transient operation, providing sufficient boost pressure in the presence of high EGR, controlling HC and CO given the lower combustion and hence exhaust temperatures, ensuring acceptable noise levels at high load, and optimization of injector and combustion chamber design and injection strategies. There do not appear to be any fundamental reasons why these problems cannot be solved or a suitable compromise reached between conflicting requirements.

A possible solution could be an engine with a CR of around 16, with direct injection at modest injection pressures (<500 bar [50 MPa]) and running on a fuel with a RON of 70 to 85 but with no strict limits on volatility. The engine could start with a spark and run in SI mode during warm-up with some throttling to ensure lower pressures in the cylinder to ease spark breakdown. Alternatively, starting could be facilitated by a sufficiently powerful glow plug. It would run in CI mode at all other conditions using multiple injections. It would have an oxidation catalyst to control CO and HC. It could also have a particulate filter that would be lightly loaded, if at all, at most loads. If soot is produced at high loads because PCI combustion is not possible or is avoided because of high-pressure rise rates, the particulate filter would be self-regenerating because the exhaust temperature would be high enough. However, such an engine would require sufficient boost pressure/inlet oxygen in the presence of high levels of EGR required to control NOx and hence might require a supercharger as well as a turbocharger.

Given that the diesel and SI engine designs are moving toward each other, it is not surprising that the best fuel solution also would be between gasoline and diesel fuels. However, the optimum fuel is likely to be much more like today's gasolines than today's diesels and could be simpler to manufacture than either. The implications for future fuels will be discussed further in Chapter 7.

6.5 References

6.1 Stone, R. 1999. Chap. 2. In *Introduction to Internal Combustion Engines*. SAE International, Warrendale PA.

6.2 Dec, J.E. 1997. "A Conceptual Model of DI Diesel Combustion Based on Laser Sheet Imaging." SAE Paper No. 970873. SAE International, Warrendale PA.

6.3 Dec, J., and Canaan, R. 1998. "PLIF Imaging of NO Formation in a DI Diesel Engine." SAE Paper No. 980147. SAE International, Warrendale PA.

6.4 Kamimoto, T., and Bae, M. 1988. "High Combustion Temperature for the Reduction of Particulate in Diesel Engines." SAE Paper No. 880423. SAE International, Warrendale PA.

6.5 Dec, J.E. 2009." Advanced Compression Ignition Engines—Understanding the in-Cylinder Processes." *Proceedings of the Combustion Institute* 32:2727–2742.

6.6 Johnson, T. 2005. "Diesel Engine Emissions and Their Control." *Platinum Metals Review* 52(1):23–37.

6.7 "Emission Standards: Summary of Worldwide Engine Emission Standards." DieselNet. http://www.dieselnet.com/standards/. Accessed August 7, 2012.

6.8 Kalghatgi, G.T., Risberg, P., and Ångström, H.-E. 2006. "Advantages of a Fuel with High Resistance to Auto-ignition in Late-Injection, Low-Temperature, Compression Ignition Combustion." SAE Paper No. 2006-01-3385. SAE International, Warrendale PA.

6.9 Kalghatgi, G.T., Risberg, P., and Ångström, H.-E. 2007. "Partially Premixed Auto-ignition of Gasoline to Attain Low Smoke and Low NOx at High Load in a Compression Ignition Engine and Comparison with a Diesel Fuel." SAE Paper No. 2007-01-006. SAE International, Warrendale PA.

6.10 Kalghatgi, G.T., L. Hildingsson, B. Johansson. 2010. "Low NO_x and Low Smoke Operation of a Diesel Engine Using Gasoline-Like Fuels." ASME Paper No. ICES2009-76034. *Journal of Engineering for Gas Turbines and Power* 132(9) doi: 10.1115/1.4000602.

6.11 Hildingsson, L., Kalghatgi, G., Tait, N., Johansson, B., and Harrison, A. 2009. "Fuel Octane Effects in the Partially Premixed Combustion Regime in Compression Ignition Engines." SAE Paper No. 2009-01-2648. SAE International, Warrendale PA.

6.12 Hildingsson, L. Johansson, B., Kalghatgi, G.T., and Harrison, A.J. 2010. "Some Effects of Fuel Autoignition Quality and Volatility in Premixed Compression Ignition Engines." SAE Paper No. 2010-01-0607. SAE International, Warrendale PA.

6.13 Kalghatgi, G.T., Hildingsson, L. Johansson, B., and Harrison, A.J. 2011. "Autoignition Quality of Gasolines in Partially Premixed Combustion in Diesel Engines." *Proceedings of the Combustion Institute* 33:3015–3021.

6.14 Kalghatgi, G.T., Hildingsson, L. Johansson, B., and Harrison, A.J. 2010. "Low-NOx, Low-Smoke Operation of a Diesel Engine Using "Premixed Enough" Compression Ignition—Effects of Fuel Autoignition Quality, Volatility and Aromatic Content." *THIESEL 2010, Thermo and Fluid Dynamic Processes in Diesel Engines, September 14–17, Valencia*, pp. 409–420.

6.15 Kalghatgi, G.T., Hildingsson, L., Harrison, A.J., L., and Johansson, B. 2011. "Surrogate Fuels for Premixed Combustion in Compression Ignition Engines." *International Journal of Engine Research* 12(5):452–465.

6.16 Kalghatgi, G.T., Gurubaran, K., Davenport, A., Taylor, A.M., Hardalupas, Y. 2013. "Some Advantages and Challenges of Running a Euro IV, V6 Diesel Engine on a Gasoline Fuel," *Fuel* 108:197–207.

6.17 Kalghatgi, G.T. 2005. "Auto-Ignition Quality of Practical Fuels and Implications for Fuel Requirements of Future SI and HCCI Engines." SAE Paper No. 2005-01-0239. SAE International, Warrendale, PA.

6.18 Heywood, J.B. 1988. Chaps. 10 and 11. In *Internal combustion engine fundamentals*. McGraw-Hill, New York.

6.19 Xue, H., and Aggarwal, S.K. 2003. "NOx Emissions in N-Heptane/Air Partially Premixed Flames." *Combustion and Flame* 132:723–741.

6.20 Colban, W.F., Miles, P.C., and Oh, S. 2007. "On the Cyclic Variability and Sources of Unburned Hydrocarbon Emissions in Low Temperature Diesel Combustion Systems." SAE Paper No. 2007-01-1837. SAE International, Warrendale, PA.

6.21 Opat, R., Ra, Y., Manuel, A., Gonzalez, D., Krieger, R., Reitz, R.D., Foster, D.E., Durrett, R.P., and Siewert, R.M. 2007. "Investigation of Mixing and Temperature Effects on HC/CO Emissions for Highly Dilute Low Temperature Combustion in a Light Duty Diesel Engine." SAE Paper No. 2007-01-0193. SAE International, Warrendale, PA.

6.22 Hochhauser, A.M. 2008. CRC Report E-84, "Review of Prior Studies of Fuel Effects on Vehicle Emissions." Coordinating Research Council, August 2008. http://www.crcao.com/reports/recentstudies2008/E--84/E-84%20Report%20Final,%20Aug%2014.pdf. Accessed January 28, 2013. Also published as "Review of Prior Studies of Fuel Effects on Vehicle Emissions." SAE Paper No. 2009-01-1181. SAE International, Warrendale, PA.

6.23 Farnell, R., and Riding, D. 2003. "Engine Noise," Chapter 20. In *Diesel Engine Reference Book*, edited by B. Challen and R. Baranescu, Ellesvier Butterworth-Heinemann, Oxford.

6.24 Reinhart, T.E. 1987. "An Evaluation of Lucas Combustion Noise Meter on Cummins B Series Engines." SAE Paper No. 870952. SAE International, Warrendale, PA.

6.25 Chan, C.M.P., Moncrieff, I.D., and Pettitt, R.A. 1982. "Diesel Engine Combustion Noise with Alternative Fuels." SAE Paper No. 820236. SAE International, Warrendale, PA.

6.26 Zhang, L. 1999. "A Study of Pilot Injection in a DI Diesel Engine." SAE Paper No. 1999-01-3493. SAE International, Warrendale, PA.

6.27 Minami, T., Takeuchi, K., and Shimazaki, N. 1995. "Reduction of Diesel Engine NOx Using Pilot Injection." SAE Paper No. 950611. SAE International, Warrendale, PA.

6.28 Ickes, A.M., Bohac, S.V., and Assanis, D.N. 2009. "Effect of Fuel Cetane Number on a Premixed Diesel Combustion Mode." *International Journal of Engine Research* 10:251.

6.29 Risberg, P., Kalghatgi, G.T., and Ångström, H.-E. 2005. "Autoignition Quality of Diesel-Like Fuels in HCCI Engines." SAE Paper No 2005-01-2127. SAE International, Warrendale, PA.

6.30 Li, T., Moriwaki, R., Ogawa, H., Kakizaki, R., and Murase, M. 2012. "Dependence of Premixed Low-Temperature Diesel Combustion on Fuel Ignitability and Volatility." *International Journal of Engine Research* 13:14.

6.31 Schleyer, C.H., Duffy, K.P., Liechty, M.P., Hardy, W.L., and Bessonette, P.W. 2007. "Effects of Fuel Property Changes on Heavy Duty HCCI Combustion." SAE Paper No. 2007-01-0191. SAE International, Warrendale, PA.

6.32 Yun, H., Sellnau, M., Milovanovic, N., and Zuelch, S. 2008. "Development of Premixed Low-Temperature Diesel Combustion in a HSDI Diesel Engine." SAE Paper No. 2008-01-0639. SAE International, Warrendale, PA.

6.33 Siebers, D., and Higgins, B. 2001. "Flame Lift-Off on Direct-Injection Diesel Sprays Under Quiescent Conditions." SAE Paper No. 2001-01-0530. SAE International, Warrendale, PA.

6.34 Pickett, L., Siebers, D., and Idicheria, C. 2005. "Relationship between Ignition Processes and the Lift-Off Length of Diesel Fuel Jets." SAE Paper No. 2005-01-3843. SAE International, Warrendale, PA.

6.35 Mueller, C., Pitz, W., Pickett, L., Martin, G., et al. 2003. "Effects of Oxygenates on Soot Processes in DI Diesel Engines: Experiments and Numerical Simulations." SAE Paper No. 2003-01-1791. SAE International, Warrendale, PA.

6.36 Miyamoto, N., Ogawa, H., Nurun, N., Obata, K., and Arima, T. 1998. "Smokeless, Low NOx, High Thermal Efficiency, and Low Noise Diesel Combustion with Oxygenated Agents as Main Fuel." SAE Paper No. 980506. SAE International, Warrendale, PA.

6.37 Choi, C.Y., and Reitz, R.D. 1999. "An Experimental Study on the Effects of Oxygenated Fuel Blends and Multiple Injection Strategies on DI Diesel Engine Emissions." *Fuel* 78: 1303–1317.

6.38 González D., M., Piel, W., Asmus, T., Clark, W., et al. 2001. "Oxygenates Screening for Advanced Petroleum-Based Diesel Fuels: Part 2. The Effect of Oxygenate Blending Compounds on Exhaust Emissions." SAE Paper No. 2001-01-3632. SAE International, Warrendale, PA.

6.39 Sienicki, E., Jass, R., Slodowske, W., McCarthy, C., et al. 1990. "Diesel Fuel Aromatic and Cetane Number Effects on Combustion and Emissions from a Prototype 1991 Diesel Engine." SAE Paper No. 902172. SAE International, Warrendale, PA.

6.40 Yanagihara, H., Sato, Y., and Mizuta, J. 1996. "A Simultaneous Reduction in NOx and Soot in Diesel Engines Under a New Combustion System (Uniform Bulky Combustion System—UNIBUS)." *17th International Vienna Motor Symposium*, pp. 303–314.

6.41 Hasegawa, R., and Yanagihara, H. 2003. "HCCI Combustion in DI Diesel Engine." SAE Paper No. 2003-01-0745. SAE International, Warrendale, PA.

6.42 Kimura, S., Osama, A., Ogama, H., Muranaka, S. 1999. "New Combustion Concept for Ultra-Clean and High-Efficiency Small DI Diesel Engines." SAE Paper No. 1999-01-3681. SAE International, Warrendale, PA.

6.43 Kimura, S., Osamu, A., Kitahara, Y., and Aiyoshizawa, E. 2001. "Ultra-Clean Combustion Technology Combining a Low Temperature and Premixed Combustion Concept for Meeting Future Emission Standards." SAE Paper No. 2001-01-0200. SAE International, Warrendale, PA.

6.44 Kimura, S., Ogawa, H., Matsui, Y., and Enomoto, Y. 2002. "an Experimental Analysis of Low Temperature and Premixed Combustion for Simultaneous Reduction of NOx and Particulate Emissions in Direct Injection Diesel Engines." *International Journal of Engine Research* 3/4:249–259.

6.45 Hashizume, T., Miyamoto, T., Akagawa, H., and Tsujimura, K. 1998. "Combustion and Emission Characteristics of Multiple Stage Diesel Combustion." SAE Paper No. 980505. SAE International, Warrendale, PA.

6.46 Mueller, C.J., Martin, G.C., Briggs, T.E., and Duffy, K.P. 2004. "an Experimental Investigation of in-Cylinder Processes Under Dual-Injection Conditions in a DI Diesel Engine." SAE Paper No. 2004-01-1843. SAE International, Warrendale, PA.

6.47 Dronniou, N., Lejeune, M., Balloul, I., and Higelin, P. 2005. "Combination of High EGR Rates and Multiple Injection Strategies to Reduce Pollutant Emissions." SAE Paper No. 2005-01-3726. SAE International, Warrendale, PA.

6.48 Keeler, B., and Shayler, P. 2008. "Constraints on Fuel Injection and EGR Strategies for Diesel PCCI-Type Combustion," SAE Paper No. 2008-01-1327. SAE International, Warrendale, PA.

6.49 Manente, V., Johansson, B., and Tunestål, P. 2009. "Partially Premixed Combustion at High Load using Gasoline and Ethanol, a Comparison with Diesel." SAE Paper No. 2009-01-0944. SAE International, Warrendale, PA.

6.50 Manente, V., Johansson, B., Tunestål, P., Canella, W. 2009. "Effects of Different Types of Gasoline Fuels on Heavy Duty Partially Premixed Combustion." SAE Paper No. 2009-01-2668. SAE International, Warrendale, PA.

6.51 Manente, V., Johansson, B., and Canella, W. 2011. "Gasoline Partially Premixed Combustion, the Future of Internal Combustion Engines?" *International Journal of Engine Research* 12:194–208.

6.52 Hanson, R., Splitter, D., and Reitz, R. 2009. "Operating a Heavy-Duty Direct-Injection Compression-Ignition Engine with Gasoline for Low Emissions." SAE Paper No. 2009-01-1442. SAE International, Warrendale, PA.

6.53 Cracknell, R.J., Rickeard, D.J., Ariztegui, J., Rose, K.D., Meuther, M., Lamping, M., and Kolbeck, A. 2008. "Advanced Combustion for Low Emissions and High Efficiency. Part 2: Impact of Fuel Properties on HCCI Combustion." SAE Paper No. 2008-01-2404. SAE International, Warrendale, PA.

6.54 Sellnau, M., Sinnamon, J., Hoyer, K., and Husted, H. 2011. "Gasoline Direct Injection Compression Ignition (GDCI)—Diesel-Like Efficiency with Low CO_2 Emissions." SAE Paper No. 2011-01-1386. SAE International, Warrendale, PA.

6.55 Sellnau, M.C., Sinnamon, J., Hoyer, K., Husted, H. 2012. "Full-Time Gasoline Direct-Injection Compression Ignition (GDCI) for High Efficiency and Low NOx and PM." SAE Paper No. 2012-01-0384. SAE International, Warrendale, PA.

6.56 Ra, Y., Loeper, P., Reitz, R., Andrie, M., et al. 2011. "Study of High Speed Gasoline Direct Injection Compression Ignition (GDICI) Engine Operation in the LTC Regime." *SAE International Journal of Engines* 4(1):1412–1430.

6.57 Borgqvist, P., Tunestal, P., and Johansson, B. 2012. "Gasoline Partially Premixed Combustion in a Light Duty Engine at Low Load and Idle Operating Conditions." SAE Paper 2012-01-0687. SAE International, Warrendale, PA.

6.58 Weall, A.J., and Collings, N. 2009. "Gasoline Fuelled Partially Premixed Compression Ignition in a Light Duty Multi Cylinder Engine: A Study of Low Load and Low Speed Operation." SAE Paper No. 2009-01-1791. SAE International, Warrendale, PA.

6.59 Ra, Y., Loeper, P., Andrie, M., Krieger, R., et al. 2012. "Gasoline DICI Engine Operation in the LTC Regime Using Triple-Pulse Injection." *SAE International Journal of Engines* 5(3).

6.60 Chang, J., Viollet, Y., Amer, A., and Kalghatgi, G. 2012. "Enabling High Efficiency Direct Injection Engine with Naphtha Fuel through Partially Premixed Compression Ignition Combustion." SAE Paper No. 2012-01-0677. SAE International, Warrendale, PA.

6.61 Weall, A.J., and Collings, N. 2007. "Investigation into Partially Premixed Combustion in a Light Duty Multi Cylinder Diesel Engine Fuelled with a Mixture of Gasoline and Diesel." SAE Paper No. 2007-01-4058. SAE International, Warrendale, PA.

6.62 Won, H-W, Peters, N., Tait, N., and Kalghatgi, G.T. 2012. "Sufficiently Premixed Compression Ignition of a Gasoline-Like Fuel Using Three Different Nozzles in a Diesel Engine." *Proceedings of the Institution of Mechanical Engineers Part D: Journal of Automobile Engineering* 226(5):698–708.

6.63 Won, H-W, Pitsch, H., Tait, N., and Kalghatgi, G.T. 2012 "Some Effects of Gasoline and Diesel Mixtures on Partially Premixed Combustion and Comparison with Practical Fuels, Gasoline and Diesel, in a Diesel Engine." *Proceedings of the Institution of Mechanical Engineers Part D: Journal of Automobile Engineering* 226(9):1259–1270.

6.64 Fan Zhang, F., Hongming Xu, Rezaei, S.Z., Kalghatgi, G., and Shi-jin Shuai. 2012. "Combustion and Emission Characteristics of a PPCI Engine Fuelled with Dieseline." SAE Paper No. 2012-01-1138. SAE International, Warrendale, PA.

6.65 Splitter, D., Reitz, R., and Hanson, R. 2010. "High Efficiency, Low Emissions RCCI Combustion by Use of a Fuel Additive." *SAE International Journal of Fuels and Lubricants* 3(2):742–756.

6.66 Kokjohn, S., Hanson, R., Splitter, D., Kaddatz, J., et al. 2011. "Fuel Reactivity Controlled Compression Ignition (RCCI) Combustion in Light- and Heavy-Duty Engines." *SAE International Journal of* Engines 4(1):360–374.

6.67 Splitter, D., Hanson, R., Kokjohn, S., and Reitz, R. 2011. "Reactivity Controlled Compression Ignition (RCCI) Heavy-Duty Engine Operation at Mid-and High-Loads with Conventional and Alternative Fuels." SAE Paper No. 2011-01-0363. SAE International, Warrendale, PA.

6.68 Hanson, R., Kokjohn, S., Splitter, D., and Reitz, R. 2011. "Fuel Effects on Reactivity Controlled Compression Ignition (RCCI) Combustion at Low Load." *SAE International Journal of Engines* 4(1):394–411

6.69 Kokjohn, S.L., Hanson, R.M., Splitter, D.A., and Reitz, R.D. 2011. "Fuel Reactivity Controlled Compression Ignition (RCCI): A Pathway to Controlled High-Efficiency Clean Combustion." *International Journal of Engine Research* 12:209–226.

6.70 Kaddatz, J., Andrie, M., Reitz, R., and Kokjohn, S. 2012. "Light-Duty Reactivity Controlled Compression Ignition Combustion Using a Cetane Improver." SAE Technical Paper No. 2012-01-1110. SAE International, Warrendale, PA.

6.71 Nieman, D.E., Dempsey, A.B., and Reitz, R. 2012. "Heavy-Duty RCCI Operation Using Natural Gas and Diesel." SAE Paper No. 2012-01-0379. SAE International, Warrendale, PA.

6.72 Srinivasan, K.K., Krishnan, S.R., et al. 2006. "The Advanced Injection Low Pilot Ignited Natural Gas Engine: A Combustion Analysis." *Journal of Engineering for Gas Turbines and Power* 128:213.

6.73 Dempsey, A., and Reitz, R. 2011. "Computational Optimization of a Heavy-Duty Compression Ignition Engine Fueled with Conventional Gasoline." *SAE International Journal of Engines* 4(1):338–359.

6.74 Bhave, A., Coble, A., Mossbach, S., Kraft, M., Morgan, N., and Kalghatgi, G. 2011. "Simulating PM Emissions and Combustion Stability in Gasoline/Diesel Fuelled Engines." SAE Paper No. 2011-01-1184. SAE International, Warrendale, PA.

6.75 Krishnan, S.R., and Srinivasan, K.K. 2010. "Multi-Zone Modelling of Partially Premixed Low-Temperature Combustion in Pilot-Ignited Natural-Gas Engines." *Proceedings of the Institution of Mechanical Engineers, Part D: Journal of Automobile Engineering* 224(12):1597–1622.

6.76 Morgan, N., Smallbone, A., Bhave, A., Kraft, M., Cracknell, R., and Kalghatgi, G. 2010. "Mapping Surrogate Gasoline Compositions into RON/MON Space." *Combustion and Flame* 157:1122–1113.

6.77 "DiesOtto." Wikipedia. http://en.wikipedia.org/wiki/DiesOtto. Accessed July 1, 2012.

Chapter 7
Implications for Future Transport Fuels

Around 95% of global energy for transport is now supplied by liquid fuels made from petroleum, and petroleum will continue to be the primary source of energy for transport in the future. The global annual demand for transport fuels is large, and even though the supply of alternative fuels such as biofuels is likely to grow, they have limited capacity to replace conventional petroleum-based fuels. The demand for transport, and hence transport fuels, is increasing rapidly, primarily driven by growth in non-OECD countries. However, this growth in fuel demand will be skewed very much toward commercial transport (Chapter 1). Currently, the passenger car sector is dominated by spark ignition (SI) engines using gasoline, whereas the commercial road transport and marine sectors are dominated by diesel engines using diesel fuel. If this state of affairs is not changed, the demand for diesel fuel is expected to increase by 85%, while the demand for gasoline is expected to decrease by 15% by 2040 compared to 2010 in some projections [7.1]. For future fuel demand and supply to be in balance, investment will be required in fuels manufacture. At the same time, future compression ignition engines will need to substantially use fuels other than current diesel fuels, for example, poor-quality gasoline such as naphtha, CNG (compressed natural gas), and GTL (gas-to-liquid) fuels.

As discussed in Chapters 4, 5, and 6, engine technology is also changing, and this will affect the fuel requirements of future engines. The efficiency is lower [7.2], and therefore CO_2 emission per mile traveled is higher for an SI engine than a diesel engine of the same capacity. However, diesel engines also produce NOx and particulate matter (PM) because the fuel and air are not well mixed prior to combustion. Controlling these diesel emissions is becoming increasingly important but is very difficult. Aftertreatment technologies that have been developed add extra cost to the diesel engine, which is already much more expensive compared to an SI engine of similar capacity, and these technologies could reduce the engine's efficiency. The primary development trend for SI engines is aimed at improving their efficiency, whereas for diesel engines the aim is to reduce their soot and NOx emissions without compromising their efficiency, while controlling cost.

7.1 Fuel Implications of Spark Ignition Engine Development Trends

As discussed in Chapter 4, fuel efficiency is always limited by knock in SI engines, and any attempt to increase efficiency makes knock more likely. Higher efficiency is enabled by higher antiknock quality of the fuel, which is described by the octane index, $OI = (1 - K)\text{RON} + K\text{MON} = \text{RON} - KS$, where RON and MON are the research and motor octane numbers (Chapter 2), K is a constant depending on the pressure and temperature history of the fuel/air mixture in the cylinder, and S is the sensitivity (RON – MON). Moreover, these efforts to improve efficiency and power density increase the pressure in the cylinder for a given mixture temperature, and nonparaffinic fuels become relatively more resistant to knock as pressure is increased for a given temperature. Primary reference fuels, on which the octane scale is based, are mixtures of two paraffins, iso-octane and n-heptane; practical fuels contain many nonparaffinic components. The result is that in modern engines for a given RON, a lower MON fuel has better antiknock quality—modern engines have negative values of K. Initiatives to improve efficiency such as downsizing and turbocharging will ensure that K is negative in such engines (see Chapter 4). Thus future fuels for SI engines need to be as different from paraffinic fuels as possible. The primary sources of increased RON, as well as sensitivity, in practical gasolines are aromatics, olefins, and oxygenates like MTBE (methyl tertiary butyl ether) and ethanol. However, there are other issues associated with ethanol, as discussed in Chapter 1. Methanol also has good antiknock properties and could become an important future component if the problems associated with its handling and distribution are resolved. Other important fuel components will be aromatics such as toluene and olefins such as di-isobutylene (iso-octene). MTBE is also an excellent fuel component of high antiknock quality but its use has been dramatically reduced in the recent past because of concerns about contamination of the water table [7.3].

7.1.1 Specifications for Antiknock Quality

The manufacture of fuels is geared toward meeting fuel specifications. These specifications first need to be necessary and, when necessary, need to be appropriate to ensure that engines get suitable fuels. In many parts of the world, fuel specifications have been set on the assumption that MON contributes to antiknock properties of the fuel. Thus, in Europe there is a minimum specification for MON of 85, while in the United States antiknock quality is specified by (RON + MON)/2, which is equivalent to assuming that the average fleet has a *K* value of +0.5. These specifications were set when MON did matter, but in modern engines with higher efficiencies, this is not the case anymore. Modern engines have negative *K* values at operating conditions where the octane requirement is high, so that for a given RON, a lower MON fuel has higher antiknock quality and this trend will continue as engines seek to get more efficient as discussed in Chapter 4. In some engines the value of *K* might be positive at conditions when knock is less likely and the octane requirement is low (e.g., at high engine speeds). If two fuels of different sensitivity meet the octane requirement of the engine at the condition when *K* is positive, the fuel with the higher sensitivity will be nearer to satisfying the higher octane requirement of the engine when *K* is negative and antiknock quality is more important, as discussed in Chapter 4.

In many parts of the world, gasoline antiknock quality specifications are inconsistent with the requirements of current and future engines at conditions where knock is more likely, and more effort and energy are being spent in making fuels that are arguably less suitable for modern engines. For instance, in Europe if the MON specification were reduced to, say, 84 (with RON at 95) or even abolished, refineries might become more flexible and use less energy, and the gasoline would have higher antiknock quality in modern and future engines, enabling them to meet more of their potential. Another example where current octane specifications might have led to inappropriate fuel components being manufactured at higher cost is in the conversion of MTBE plants. Following the banning of the use of MTBE in many parts of the world, MTBE plants have been converted to make iso-octane [7.4], which has a MON of 100. However, iso-octene or di-isobutylene, which is an intermediate product in this process, would be cheaper to manufacture and has higher blending RON but lower blending MON (higher sensitivity) compared to iso-octane. Hence, it would have better antiknock properties in modern engines compared to iso-octane. Compared to MTBE, iso-octene would also have higher energy content and would not increase RVP (Reid vapor pressure; see Chapter 2). There are some concerns that olefinic fuels like iso-octene might cause problems with stability or injector deposits (see Chapters 2 and 3). However, these concerns are unwarranted [7.5, 7.6] for linear, nonconjugated olefins, but would need to be properly assessed if iso-octene were to be used as a major gasoline component.

Currently, there is not much appetite to change the way fuel antiknock quality is specified or to change the specification standards that have been set, because of the enormous effort needed to find consensus among several stakeholders, including legislative bodies. Vehicle manufacturers deal with the occasional shortfall of fuel antiknock quality through the use of the knock sensor and by tolerating some loss of power and performance to avoid knock. Fuel manufacturers have already made investments in their refineries to meet demands on MON (e.g., by building alkylation plants). However, as engines seek to improve their efficiency, this mismatch between specifications and engine requirements will only get wider and will need to be addressed.

It is interesting to speculate how this problem might be solved in the future. Remember that the minimum quality requirement that is set (e.g. the lower limits on RON and MON) and how the fuel antiknock quality is described (e.g., the *K* value) are both important. One approach might be to assume a *K* value of, say, −0.5 for the average fleet and then set a standard on OI. Thus the standard might be revised to say that a minimum OI of 100 is required for a gasoline with OI defined as (1.5RON − 0.5MON). The current European fuel of 95 RON and 85 MON would meet this requirement, but so would a fuel with RON of 94.5 and MON of 83.5 which would contain higher fractions of more sensitive components like olefins, aromatics and oxygenates. There are many parts of the world, such as Japan and China, where fuel antiknock quality is only specified by RON(i.e., *K* is assumed to be zero), and indeed this might be an acceptable intermediate compromise. Certainly, in such cases setting an additional MON specification would be a retrogressive step. The question also arises as to whether the minimum OI requirement should be increased and to what level. The gain in knock-limited power, knock-limited fuel efficiency, and the potential for improving efficiency through downsizing do not increase indefinitely with increasing OI [7.7]—there are diminishing returns with increasing OI. Increasing fuel antiknock quality could also require greater use of energy in making the fuel. A proper well-to-wheel analysis would be required to assess the true benefits for energy consumption attainable through increasing fuel antiknock quality in combination with downsizing.

7.1.2 Other Fuel Specifications

Other current fuel specifications make it more difficult to manufacture fuels of high antiknock quality from refinery components. For instance, aromatic levels are limited to 35% by volume in European and U.S. gasolines (Chapter 2). Since aromatics are the source of high RON as well as high sensitivity, this makes it difficult to attain high antiknock quality. These restrictions were brought in to reduce benzene emissions after auto–oil joint industry tests in the 1980s and 90s (see Chapter 2). However, engines and catalysts have improved significantly since then, and tailpipe emissions from modern engines are not very sensitive to fuel aromatic content. If some of these specifications were relaxed, it is likely that there will be reductions in energy consumption and CO_2 emissions during the manufacture of fuels of high antiknock quality because of increased flexibility in refineries. As engines and catalysts develop, fuel effects on emissions and efficiency will also change. It is important to have rational fuel specifications that also take into account the energy and CO_2 costs of manufacturing the fuels. For instance, it is now clear that fuel sulfur and benzene levels, deposit formation, and volatility characteristics are important and need to be controlled, but antiknock quality needs to be reassessed and specifications on aromatics might need to be reconsidered because of the changes in engines and aftertreatment systems that have taken place. Continuing discussion is necessary among auto and oil industries, regulatory bodies, and other stakeholders to ensure that future fuels meet the requirements of future engines while minimizing waste in fuels manufacture.

7.2 Fuel Implications of Compression Ignition Engine Development Trends

The primary development trend in compression ignition (CI) engines is to reduce NOx and smoke without compromising the efficiency and increasing the cost. HCCI (homogeneous charge compression ignition) combustion (Chapter 4) can achieve this result at very low loads, but even then HCCI is difficult to control because of the lack of in-cycle control on combustion phasing. Therefore, HCCI engines are not likely to be appropriate for practical transport applications, and the only practical CI engine now available is the diesel engine using conventional diesel fuels (CN 40–60). Simultaneous reduction of NOx and smoke is very difficult for diesel engines if a conventional diesel fuel with low ignition delay is used. As discussed in Chapter 6, the GCI (gasoline compression ignition) approach, running diesel engines on gasoline-like fuels, offers the possibility of a relatively cheap but efficient engine and aftertreatment system with acceptable emissions. Moreover, the fuel needed could have much lower octane (70–85 RON) than today's gasolines and could contain components in the diesel boiling range. Indeed, straight run gasoline, which could be easily available in the future because of the expected skewing of future demand toward commercial transport, would have octane numbers near this range and could be used with little or no upgrading in such engines.

In the short term, say, up to 2020, diesel fuel ($CN > 40$) will continue to be used in CI engines because there has been so much investment made by all the OEMs in developing the engine and aftertreatment technology for such systems. However, they will try to use PCI (premixed compression ignition) combustion as much as possible and will have to use higher injection pressures (see Chapter 6). Such systems will continue to use non-premixed combustion at high load. They will also have to use extensive aftertreatment to meet increasingly stringent emissions standards on NOx and soot. This will increase their cost and might reduce their efficiency (e.g., because of the extra fuel to be used in the regeneration of diesel particulate filters or because of lower compression ratios used to increase ignition delays). In such cases high cetane might show an advantage in terms of lower noise at low load because it makes diffusive combustion more likely. Higher cetane will also ease cold start which is compromised by lower compression ratio. Also, GTL fuels, which have no sulfur and no aromatics, would have an advantage in terms of the particulate emissions that are inevitable with non-premixed combustion.

In the long term, say after 2030, CI engines could run on poor-quality gasoline with RON in the range of 70–85 and no stringent demands on volatility. It is difficult to predict the transition between the short-term and long-term scenarios, since this will depend on the development strategies employed by the OEMs. However, GCI engines initially will have to run on existing fuels. If a single-fuel GCI strategy is used, a suitable fuel for this concept could be 90% European premium gasoline (95 RON) + 10% European diesel (52 CN), though systems running entirely on market gasoline, for example, U.S. regular with (RON + MON)/2 of 87, are being developed (Chapter 6). Even in the dual-fuel reactivity controlled compression ignition (RCCI) concept (Chapter 6), around 90% of the fuel used will be gasoline. Both these concepts, where both gasoline and diesel are upgraded in the refinery and then downgraded on the vehicle, are not sustainable in the long term. Eventually, appropriate, less processed fuels that require less energy to produce will have to be manufactured and distributed by the fuel manufacturers for such engines.

7.3 Summary and Concluding Remarks

As SI engines are developed to become more efficient, knock will become more likely. Fuels of high antiknock quality, and with high RON and preferably low MON, will enable future SI engines to reach their full potential. In many parts of the world, fuel antiknock specifications are set on the assumption that higher MON contributes to increased knock resistance. Such specifications are inconsistent with the fuel requirements of modern engines and make it more difficult to satisfy the fuel requirements of future engines. The specifications will need to be revised, as this mismatch between specifications and requirements widens. In countries where there are no MON specifications, bringing them in would be a retrogressive step. A new rating method to rank gasoline fuels for autoignition quality could be developed, as discussed in Section 4.4. Other fuel specifications, such as the limits on aromatics, make it difficult to produce fuels of high antiknock quality using refinery components. Some of these specifications, which were brought in to control tailpipe emissions, might not be needed in modern engines with modern aftertreatment systems. There should be continuing dialogue between various stakeholders to ensure the development of rational fuel specifications appropriate for modern engines while avoiding unnecessary expenditure of energy in fuel processing to meet unnecessary specifications. For instance, existing gasoline specifications on sulfur, benzene, volatility, and deposit control are necessary and important, but other specifications might need to be revised. Alternative fuels like ethanol have a big role to play in increasing the antiknock quality of future fuels. However, ethanol could also increase the likelihood of preignition in downsized and turbocharged engines. Given that increasing antiknock quality brings diminishing returns beyond a limit, say, RON of 103, it should be possible to have ethanol containing fuels with high enough antiknock quality while minimizing the chances of preignition. In fact, the chances of preignition causing super-knock might be reduced because of the high resistance to autoignition of ethanol, though it might make preignition more likely. These are areas that need further investigation. In addition, other requirements such as controlling deposits will continue to be important particularly with increasing use of direct injection (see Chapter 3). Controlling injector deposits in direct injection spark ignition (DISI) engines presents different challenges compared to older port fuel injection engines (PFI). In the future, gains in efficiency and performance from fuel formulation that might have been considered too small to bother with in the past are likely to be important as the SI engine is squeezed to the maximum. Much work is needed in quantifying such small gains and optimizing fuels for future SI engines. There is also work to be done in assessing the overall benefits of any changes to gasoline manufacture and composition on a well-to-wheel basis.

In the short term, diesel engines will continue to use conventional diesel fuels because of the huge investments made in conventional diesel engine and catalyst technology by most engine manufacturers. However, in the long term, CI engines are likely to use fuels in the gasoline autoignition range (RON > 60) but with RON lower than today's gasolines and with no strict requirements on volatility. Such fuels would require less processing in the refinery than today's fuels and would be easier to make. Moreover, light fractions like naphtha, which are expected to be in relative abundance in the future can be used with little further processing to make these fuels. Such an engine–fuels system will be at least as efficient as today's diesel engine, but will be significantly cheaper. There might be fewer barriers to the development of such GCI engines in countries like China where the commitment to and investment in traditional diesel technology is weaker.

One could imagine a long-term future where a majority (60%?) of engines are GCI engines, because of their high efficiency, that run on gasoline fuels of lower RON (70–80) and lower volatility compared to today's gasolines. The rest of the engines will be SI engines requiring high RON and preferably low MON gasoline; ethanol would be very important in making such fuels.

Other components like iso-octene (di-isobutylene) and even methanol, if the issues surrounding its practical use are resolved, could be important. More oil will be needed to supply the increased global demand for transport energy, and hence more refinery capacity will be needed. However, in this scenario, these future refineries need to be less complex than today's average refinery. Naphtha, which is likely to be easily available because of the skewing of the global demand growth toward commercial transport, can be used with little upgrading in terms of octane in GCI engines. The heavier ends of the crude oil need to be cracked to bring them into the diesel boiling range but would require less upgrading in terms of octane or cetane compared to today's practices and could be used in GCI engines. Common components would be used for making fuels for both SI and GCI engines making the refinery more flexible. Biodiesel, with current levels of cetane of over 50, would have a very limited role in such a scenario. During the transition to the long-term scenario, GCI engines will need to use existing fuels either as mixtures of gasoline and diesel or in bridging technologies such as RCCI (see Chapter 6).

In the medium term, a combination of initiatives and actions, no doubt influenced by the market, will be required to ensure that transport fuel demand is in balance with fuels manufacture. Investments will have to be made in refineries globally to shift the balance of products toward heavier fuels like diesel and jet fuel, even as diesel engines using conventional diesel fuel become more expensive and complicated to meet increasingly stringent NOx and soot requirements. At the same time, efficient CI engines will need to move from using conventional diesel fuels to use alternatives like poor-quality gasoline, CNG, and GTL. Thus changes in transport energy demand and engine technology, driven by the need to increase the efficiency of the SI engine and reduce NOx and soot from diesel engines, are likely to have a profound effect on the properties, specifications, and production of future fuels.

7.4 References

7.1 "2012 The Energy Outlook for Energy: A View to 2040." ExxonMobil. http://www.exxonmobil.co.uk/corporate/files/news_pub_eo2012.pdf. Accessed June 11, 2013.

7.2 Stone, R. 1999. Chapter 2. In *Introduction to Internal Combustion Engines.* SAE International, Warrendale PA.

7.3 "Methyl Tertiary Butyl Ether (MTBE): Drinking Water." U.S. EPA, http://www.epa.gov/mtbe/water.htm. Accessed August 13, 2012.

7.4 "MTBE Units Expansion/Conversion." CDTECH. http://www.cdtech.com/updates/Publications/Refining%20Papers/MTBE%20Unit%20Expansion-Conversion.pdf. Accessed August 13, 2012.

7.5 Pereira, R.C.C., and Pasa, V.M.D. 2006. "Effect of Mono-Olefins and Diolefins on the Stability of Automotive Gasoline." *Fuel* 85:1860–1865.

7.6 Pentkainen, J., Rantanen, L., and Aakko, P. 2004. "Effect of Heavy Olefins and Ethanol on Gasoline Emissions." SAE Paper No. 2004-01-2003. SAE International, Warrendale, PA.

7.7 Amer, A., Babiker, H., Chang, J., Kalghatgi, G., Adomeit, P., Brassat, A., and Guenther, M. 2012. "Fuel Effects on Knock in a Highly Boosted Direct Injection Spark Ignition Engine." SAE Paper No. 2012-01-1634. SAE International, Warrendale PA.

Abbreviations

A	Arrhenius A factor
$(a + b)$	slope of CA50 versus OI line
a	acoustic speed
a, b, c	regression constants linking CA50 to RON and MON
AFR	air/fuel ratio
ATDC	after top dead center
BTDC	before top dead center
CAD	crank angle degrees
CAx	the crank angle degree at which x% of total heat release has taken place
CD	combustion delay, CD = CA50 – SOI
CI	compression ignition
CN	cetane number
CO	carbon monoxide
COV	coefficient of variation = 100 (standard deviation/mean)
CR	compression ratio
DCN	derived cetane number
DD	developing detonation
E	activation energy
EGR	exhaust gas recirculation (inlet CO_2 as a percentage of exhaust CO_2)
EGR_f	exhaust gas recirculation mass fraction
EOI	end of injection
FSN	filter smoke number
GCI	gasoline compression ignition
HC	hydrocarbons
HCCI	homogeneous charge compression ignition
HRR	heat release rate, J/CAD
I	Livengood-Wu integral (Eq. 4.1)
ID	ignition delay in a CI engine (SOC – SOI)
IDL	ignition delay limit
IDW	ignition dwell or mixing period, IDW = SOC – EOI

IMEP	indicated mean effective pressure
ISCO	indicated specific [emissions of] CO
ISFC	indicated specific fuel consumption
ISHC	indicated specific [emissions of] HC
$ISNO_x$	indicated specific [emissions of] NO_x
k	thermal conductivity
K	parameter linking OI to RON and MON
KI	knock intensity
KLSA	knock-limited spark advance
KLT	knock-limited torque
Le	Lewis number
m	exponent for effect of *P* on S_L
MBT	minimum advance for best torque
MK	Modulated Kinetics (Nissan) PCI combustion
MON	motor octane number
MPa	megapascal
MPRR	maximum pressure rise rate, bar/CAD
MULDIC	multiple stage diesel combustion
n	exponent for effect of *T* on S_L
N	engine speed, rpm
NO_x	nitrogen oxides
OI	octane index, $OI = (1 - K) \cdot RON + K \cdot MON = RON - K \cdot S$
OI_0	requirement of the engine: $CA50 = 0$ when $OI = OI_0$
ON	octane number
OR	octane requirement
P	pressure
PCI	premixed compression ignition
P_{DD}	probability of DD following autoignition
P_{in}	inlet pressure, bar absolute
P_{max}	maximum pressure, bar absolute
$P_{maxcomp}$	maximum compression pressure, bar absolute
P_{PI}	probability of preignition
PR	preignition resistance
PREDIC	premixed lean diesel combustion
PRF	primary reference fuel (mixture of iso-octane and n-heptane)

Q	heat released, J
R	universal gas constant
r	general radius
r_0	original radius of spherical hot spot
r_c	critical radius for autoignition
R_f	critical radius of hot pocket of gas to initiate a flame (Eq. 5.1)
RON	research octane number
rpm	revolutions per minute
S	fuel sensitivity = RON – MON
SI	spark ignition
S_L	laminar burning velocity
S_{L0}	laminar burning velocity at P_0 and T_0
S_{Lmax}	maximum laminar burning velocity
SOC	start of combustion
SOI	start of injection
t	time
T	temperature
T_{comp15}	compression temperature when pressure is 15 bar, K
TDC	top dead center, top of the compression stroke
T_{in}	inlet temperature, °C
$T_{maxcomp}$	maximum compression temperature, °C
TN	toluene number (%vol of toluene in a TRF)
TRF	toluene reference fuel (mixture of toluene and n-heptane)
u_a	autoignition propagation velocity, relative to unburned gas (Eq. 5.5)
UNIBUS	uniform bulky combustion system (Toyota)
β	Zeldovich number (Eq. 5.1)
δ	laminar flame thickness
ξ	resonance parameter = (a/u_a) (Eq. 5.8a)
τ_i	ignition delay in premixed fuel/air mixture
ε	reactivity parameter (Eq. 5.8b)

Index

About the Author

Gautam Kalghatgi joined Saudi Aramco in October 2010 after 31 years with Shell Research in the UK. He has a B.Tech. from I.I.T. Bombay (1972) and a Ph.D. from Bristol University (1975) in aeronautical engineering. From 1975 to 1979, he did postdoctoral research in turbulent combustion at Southampton University.

He is a Fellow of the Royal Academy of Engineering, SAE International, and the Institution of Mechanical Engineers. He is a visiting professor at Imperial College, London, and has been adjunct, part-time, and visiting professor at the KTH, Royal Institute of Technology in Stockholm, Eindhoven Technical University in The Netherlands, and the University of Sheffield in the UK. He has over 100 publications on combustion, fuels, and engine research. He is on the editorial boards of three international journals: the *International Journal of Engine Research*, *Journal of Automobile Engineering*, and *Journal of Fuels and Lubricants.*